LOST AT THAXTON

The wreckage of Norfolk & Western passenger train Number Two at Thaxton. (Photo courtesy of the Virginia Room, Roanoke Public Libraries)

LOST AT THAXTON

THE DRAMATIC TRUE STORY OF VIRGINIA'S FORGOTTEN TRAIN WRECK

MICHAEL E. JONES

THAXTON PRESS

Lost at Thaxton: The Dramatic True Story of Virginia's Forgotten Train Wreck

lostatthaxton.com

Copyright © Michael E. Jones (2013, 2022)

Second Edition

All rights reserved.

ISBN 978-0-9890046-9-5 (paperback)

ISBN 978-0-9890046-3-3 (hardcover)

ISBN 979-8-9864354-0-4 (ebook)

ISBN 978-0-9890046-7-1 (large print hardcover)

ISBN 978-0-9890046-8-8 (large print paperback)

Library of Congress Control Number: 2022911011

Thaxton Press, LLC, Moore, SC

thaxtonpress.com

For information about bulk purchase discounts, please e-mail sales@thaxtonpress.com.

Cover design by Mason Jones

Front cover photograph courtesy of Norfolk and Western Historical Photograph Collection, Norfolk Southern Archives, Norfolk, Virginia. Digital image courtesy of Special Collections, Virginia Tech, Blacksburg, Virginia.

Back cover photographs courtesy of Delaware Public Archives.

No part of this book may be reproduced in any form or by any electronic or mechanical means, including information storage and retrieval systems, without written permission from the author, except for the use of brief quotations in a book review.

For Mom and Dad,

*Your love for home inspired this book,
and your love for me made it possible.*

CONTENTS

Preface ix
Introduction xiii

PART I
1. Rain 3
2. Departure 13
3. Souls 19
4. Forewarning 35
5. Disaster 41
6. Conflagration 59
7. Aftermath 69

Wreck from Southern Side 91
Wreck from Northern Side 92
Passenger Sketch of the Wreck 93
Doctors at the Wreck 94
Hotel Roanoke 95
Cleveland Monument 96
Engine Number 30 Bell Clapper 97
Culvert at Wolf Creek 98
Culvert and Fill at Wolf Creek 99
Location of Thaxton Station 100
Thaxton Milepost Marker 101
The Peaks of Otter 102

PART II
8. Blame 105
9. Who Is John Dowell? 119
10. Down the Line 125
11. Sacred Dust 131
12. Biography 137

Acknowledgments 193
Notes 199
Bibliography 205
About the Author 209

PREFACE

On the wall of her old farmhouse in Thaxton, Virginia, my Grandma Jones had hung an antique oval frame which held a timeworn photo of my great-great grandfather, Tandy Jones. Tandy was in his later years when the photo was taken, his full white beard conspicuous against a coal-colored coat. Whenever I look at old photos from that era and before, I often imagine the interesting stories that must have swirled around that person. But as a young boy, I kept my attention on those steely-blue eyes staring back at me from that antique frame. Tandy's eyes followed me no matter where I went in the room. You might chalk that up to the active imagination of a child, but I'll bet you'd feel the same after staring at those eyes for just a few moments. If you were like me, you would at least pause before getting into any mischief while you were in the vicinity of that watchful gaze.

Beyond Tandy's relentless watch over Grandma's sitting area, I did not know much about the man in the photo. Grandma Jones had told us that Tandy was a railroad section master for Norfolk & Western in

Thaxton, whatever that meant. Little did I know that my great-great grandfather's chosen occupation would lead me to spend the better part of two years researching and writing a story about a night on the job that he probably never forgot.

My involvement in Tandy's story started one summer evening in 2011 while on a beach trip with family. The discussion turned to history, as it often does if I get a chance to steer it that way, and at some point, someone mentioned a terrible train wreck that had occurred at Thaxton long ago. Even though Tandy oversaw the section of rail where the accident took place, I had never heard any mention of the wreck before.

I grew up in South Carolina, but my dad was born and raised in Thaxton, and my mom in Montvale, about seven miles west of there. Montvale and Thaxton are part of Bedford County, and nearly all my relatives have lived in the area at one point or another. When I was growing up, just about every holiday road trip our family took was to Bedford County. When my parents retired, they left South Carolina and moved back to Thaxton. Mom and Dad loved Bedford County, and they would usually refer to it as "home," no matter where they were living.

Over the years, I had spent many hours sitting on my grandmother's porch on the mountain, listening to family share old-time stories. I thought I had heard them all. When I found out about the horrific train wreck at Thaxton, I was amazed that I hadn't heard about it before. I had no idea that the wreck transpired only a handful of miles from my grandma's house. I wanted to know more about the story.

As I began to dig into the history of the wreck, I was surprised to find that there was significant loss of life, and the details of the accident were unbelievably terrifying and heartbreaking. Yet there seemed to be no memorializing of the wreck or of those who lost their lives that night in 1889. The research that I plunged into later revealed to me that this lack of attention was not out of the ordinary. There were many railroad wrecks during those early days of rapid expansion. Railroad companies consistently implemented improvements in technology to make rail travel safer, but accidents just seemed to be part of the package. Some wrecks were remembered in famous songs like "The Ballad of

Casey Jones" or "The Wreck of the Old 97," but such memorials were hit or miss. It often seemed that the personality of the engineer or the lore surrounding a wrecked train determined whether it would be remembered. The number of casualties in a wreck did not necessarily have any bearing on how well it was memorialized.

The wreck of passenger train Number Two at Thaxton seemed to slip away completely from the pages of history. Historical markers in Virginia are located at the sites of the "Wreck of the Old 97" in Danville and the "Wreck at the Fat Nancy" in Orange. The markers describe those wrecks as two of the worst in Virginia history, although both paled in comparison to the number of lives lost at Thaxton. As you will read in this book, the events that took place after the wreck that night raised the horror of this deadly accident to unimaginable heights.

This book was written to give a fitting memorial not only to those lives that were lost, but also to those who lived on and carried the scars of this tragedy with them in one way or another. It is common for us to focus on the details of the wreck and the train itself, but often we lose sight of the actual people who were part of it. I wanted to honor the people involved in the wreck at Thaxton, and I reviewed over four hundred individual sources of information to compose their story. Those sources included historical newspapers, books, court documents, personal letters, state and federal records, and personal interviews with descendants of passengers and families who live in the Thaxton area. Each of the seventy-four known passengers and crew on the train that night were researched individually in order to provide the most accurate telling possible of a story long forgotten.

INTRODUCTION

Lost at Thaxton is divided into two main sections. Part I, "Their Story," is written in the form of a story-based narrative history. I want you to be able to feel as if you were riding along on the train that night and that you were an eyewitness to all the events that took place. Even though the story is told in this manner, every detail that you read is directly related to information found through research. Descriptions of the weather, the geography, physical descriptions of passengers and crew, and even descriptions of the train itself are all based on eyewitness statements and historical documents.

Conversations and statements made by passengers and crew that night quoted during the course of the story are the actual words documented at the time. No additional dialogue was added if there was no evidence that it took place. In other words, this is not a book "based on a true story," but instead it is an authentic true story. The real-life drama that took place that night stands on its own. That is certainly not to say the book is infallible, but it is a genuine effort to tell as true a tale as can be told while looking back so long ago.

Part II is titled "The Rest of the Story," and it provides a deeper look into the causes of the wreck, what happened in the months and years afterward, and some mysteries that were revealed during the research for the book. In the final chapter, you are introduced to each person on the train. Any photographs that were uncovered of the passengers and crew can be found there.

History is at the mercy of the hands that document it, and that fact has not been taken lightly in the writing of this book.

MJ

PART I

THEIR STORY

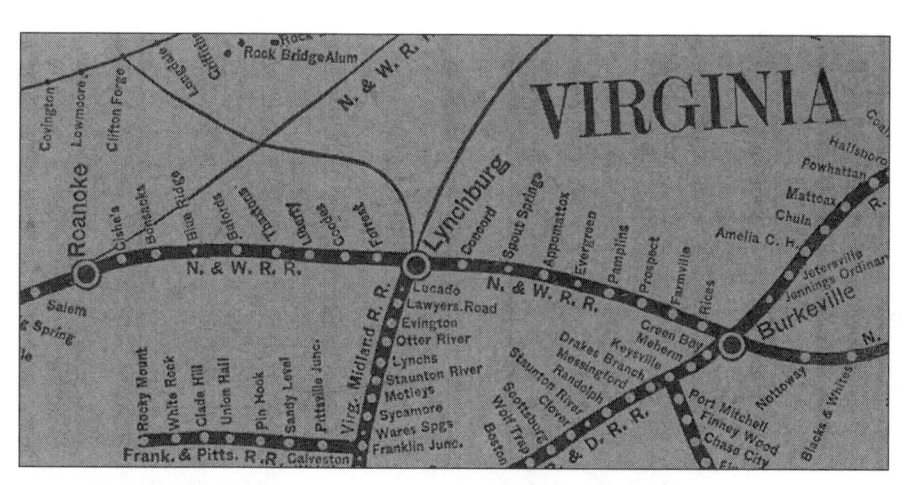

Route of Norfolk & Western passenger train Number Two. Scheduled for departure east from Roanoke, Virginia on July 1, 1889 at 11:56 p.m. Expected arrival at Thaxton, Virginia was 12:40 a.m., July 2, 1889. (Courtesy of Library of Congress, Geography and Map Division)

1

RAIN

"I have never seen anything in the vicinity equal the recent floods occurring on June 30th, and July 1st, 1889."[1]

— GEORGE A. NICHOLS, 52-YEAR-OLD NATIVE OF
BEDFORD COUNTY, JULY 12, 1889

Mother Nature had sprung a leak. As best as anyone could recall, 1889 was the wettest year ever in the state of Virginia, and for that matter all over the United States. Flash floods, or "freshets," seemed to make the news nearly every day because of a hard rain blasting its way down from the heavens on some poor, defenseless town.

The soaking was especially heavy during the spring and summer months. Richmond, Virginia's capital city, had dealt with severe flooding throughout May and June. Bridges and railroads had been battered by the weather, dams were broken, and property damage in general was widespread throughout the state. The *Richmond Dispatch* recorded over thirty-eight inches of rain through the first six months of 1889, the amount the former capital of the Confederacy would typically see for an entire year.[2]

The watery weather stayed true to form over the weekend of June 29 and 30 in the summer of 1889. Storms swept across the United States and drenched everything along the way. Texas was so saturated that rivers all over the Lone Star State were in flood stage. Several children had drowned in the swollen rivers, and railroad travel was nearly at a complete standstill. The storms that moved through that weekend caused damages estimated at more than one million dollars in the Fort Worth area alone.

On Sunday, June 30, the *Dallas Morning News* noted that "the heaviest rain perhaps that ever visited this portion of country fell here yesterday, commencing about 2 o'clock PM continuing from that time up till 12 o'clock today with very little intermission."[3] In Waco on Monday morning, July 1, a water spout struck town and caused the rain to fall "with such force that livestock fled in terror and weak animals were beat to the earth."[4] A creek in Waco rose six feet in just five minutes because of the torrential rains. It was astounding how quickly a heavy rain could swell a small creek into a raging river.

Rainstorms that weekend pummeled the country all the way from Texas to Virginia and even north to Johnstown, Pennsylvania, where the weather had been a merciless villain already that year. On May 31, heavy rains caused a faulty dam above Johnstown to succumb to the pressure of the bloated 450-acre lake behind it, and the valley below the dam was flooded. The flood killed over 2,200 men, women, and children in what was described as a "Valley of Death."[5]

The waterlogged changeover from June to July simply brought no signs that the end of the misery was at hand. The weekend storms flooded the few houses that had somehow remained standing in Johnstown, carrying off precious furniture that some folks had managed to salvage from the terrible flood.

It was the kind of year that could make you think the covenant with Noah had been broken, and the whole world might just float away.

On Monday morning, July 1, a brief moment of relief had arrived. Only a slight rain was falling over a small village called Thaxton, located at the foot of the Blue Ridge Mountains in central Virginia. Gray skies were not capable of blotting out the astounding natural beauty of the area, with its green rolling hills, fertile farmland, and the majestic Peaks of Otter rising high above the valley. Sharp Top and Flat Top, two of the three summits that comprised the Peaks of Otter, were so massive that Thomas Jefferson once believed they might be the tallest mountains in North America.[6]

Tandy Jones was not in the mood to sit back and take in the local scenery that morning. He was hard at work with one of his railroad section hands, Anderson Holt. It had been an all-night affair for Tandy and his crew, and they were still clearing out debris the storms had washed down the creeks and streams the night before. Tandy was a section master for the Norfolk & Western Railroad, and his section encompassed the area around the small village of Thaxton in Bedford County. The community was named after David Thaxton, the landowner who deeded a right of way in the 1850s to the old Virginia & Tennessee Railroad. A station and sidetrack for allowing trains to pass each other were constructed, and, like many towns, steam engines breathed Thaxton into existence. Sidetracks were commonly called "sidings", and the town was often referred to as "Thaxton's Switch"[7] because of the railroad siding. In addition to the siding, a post office was established in the town. The railroad provided employment to many of the men in the area, including Section Master Jones.

A section master's job required long hours, persistence, and attention to detail. Along with keeping the track and culverts clear, he was also responsible for ensuring the track lined up properly. The settling of the earth and the impact of heavy trains would jar the track out of alignment. An uneven track could cause the cars to jerk from side to side as they traveled over the rails, and the additional wear and tear damaged both the cars and the rails. Passengers were not too fond of lurching cars either. Some section masters would occasionally ignore culverts because they were overly focused on having the "best riding section on the road."[8] Ignoring the culverts could be a costly mistake and would result in damage to both life and property. Small items could wash down streams and create obstructions that would then trap larger debris in a culvert carrying a stream beneath an embankment. Once all that debris clogged a culvert, the water would dam up on one side of the embankment. The extreme water pressure that resulted from a clogged culvert would eventually wash the entire embankment away and the track along with it.

A good section master would typically clear all the debris near the culvert and for a considerable distance up the creek, so that nothing could wash down into the culvert later. By all accounts, fifty-seven-year-old Tandy Jones was one of the best in his division. He had overseen the Thaxton section for over thirteen years without incident, and the culverts in his section had handled all the rainstorms thrown their way for thirty-five years.

Weather set the schedule for Jones and his men just as often as the railroad officials did. The ferocious storms on June 30 had forced the section master and two of his men to battle the elements all night to keep the track and culverts clear. Even a lifelong resident like George Nichols had never experienced anything like the rain that had slammed Bedford County so hard and fast. A small branch off Little Otter Creek near Nichols's home had bulged from a shallow tributary to a level deep enough to "swim a horse" after thirty minutes of rainfall. George had been around for "the big flood" that hit the area back in the 1870s, but what he saw that weekend in 1889 was considerably worse.[9]

One of the tensest moments during the night had been the scheduled nightly run of passenger train Number Two, on its way east through Thaxton toward the city of Lynchburg. Prior to the arrival of the train, Section Master Jones had instructed another of his laborers, William Preston, to go to the western edge of their section and ensure the track and culverts were clear for the train to pass through. As he walked the tracks, Preston made certain there were no trouble spots that might endanger the train and its passengers. On that night, train Number Two had passed safely, and the men continued their work maintaining the track well into Monday.

Around three o'clock in the afternoon on July 1, it seemed like the worst of the weather had passed by when Tandy Jones made his final inspection of a small culvert at Wolf Creek. The creek passed through a culvert beneath a twenty-three-foot high embankment some of the locals called "Newman's Fill."[10] One of the landowners near the creek was Thomas Newman, a forty-nine-year-old engineer for Norfolk & Western. He and his son were at work in their fields near the culvert that afternoon when they noticed Tandy making his inspection. The culvert was clear. Only four or five inches of water drifted quietly by in Wolf Creek. The shallow little creek appeared serene, like it was flowing through a painting of the Blue Ridge Mountains. It seemed to be hardly any threat to the huge embankment.

By six o'clock that evening, Tandy and his men had had an extremely long twenty-four hours. A slight rain moved through Thaxton about that time, but it cleared out almost as quickly as it arrived. Tandy was very tired, and because he was confident that all culverts were perfectly clear, he decided to call it a day. He assigned one of his men, John Johnson, to keep watch on the section during the night. John had been on duty the previous night as well, and a day's rest had prepared him to work the night shift again. Tandy headed home, and John was prepared to be at the ready if there were any unexpected surprises. Sometime between nine and ten o'clock that night, Tandy stepped out to survey the skies and check the weather one final time for the evening.

When the stars were out in Thaxton, they were clear and bright, like lanterns made of sparkling crystal hanging from the porch of a not too distant neighbor. Although they could be stunning to look at, Tandy was not thinking about stargazing. Those stars meant the rain had cleared out, and he was going to get some much-needed rest. The section master reclined on his bed and fell asleep soundly, something he rarely did.

Not long after Tandy drifted off to sleep, rain visited Thaxton again. Most of the people around town noticed the rain getting a little heavy about ten o'clock. About an hour later, something more than an ordinary rainstorm struck the area. Charles Marshall lived near a 350-foot-high ridge above the town, roughly a mile north of the railroad crossing over Newman's Fill. In the midst of all the rain, he began to hear "a mighty noise, a rumbling and roaring noise, sounding like the grinding of mill stones."[11] His father, John Marshall, lived about a half mile away from his son, and he heard a roaring that sounded "very much like an earthquake"[12] coming from the direction of the ridge. Suddenly, men all over Thaxton were shaken from their sleep by the heavy rain and the loud roaring that accompanied the deluge. Lightning streaked across the sky against the backdrop of the soaring mountain peaks. Thunder rumbled through the valley, but where was that roaring sound coming from? Was it the wind? Many of the men got up and looked through their windows, expecting to see debris flying and trees bent over under heavy winds. An extremely powerful storm had obviously ambushed Thaxton, but the wind was not blowing at all, and the air was eerily still. If the wind was not the source of that mighty noise, what could it be?

C. M. Folden had a pretty good idea he knew what it was. Folden lived directly at the foot of the ridge above Thaxton, and he suspected that at least two cloudbursts had just taken place above the ridge. He heard the "sound of water rushing down in one large body, bringing trees and rocks"[13] off the ridge. Folden had never seen a storm like this in his forty-three years on earth, and there was no way he was going back to sleep anytime soon.

Whether Folden's suspicion of a cloudburst was correct or not, it was clear that a tremendous amount of water had dropped in a very short period near that ridge. The water had to go somewhere, and that would be into the four creeks below it. Big Otter, Little Otter, Wolf, and Reed creeks were deluged by the extreme volume of rain released around the ridge. The railroad crossed Reed Creek about one and a half miles west of the Thaxton station, and two branches of Wolf Creek were crossed by the track about a half mile on either side of the station. This furious storm tested the capabilities of the culverts at those crossings like never before.

Just after eleven o'clock on the evening of July 1, track watchman John Johnson left his house to inspect the condition of the railroad. He fought his way through the heavy rain to the tool house and grabbed a shovel to clear any dirt and mud that might accumulate on the track. Johnson had stopped for a moment at the Thaxton station when he heard the frightening sound of water rushing down one of the branches of Wolf Creek. His home was not far from the banks of the creek and only two hundred feet south of the track near the station. He hurried back toward his house to check on his wife and son. It had not been very long since Johnson had left to go to work, but the creek had risen quickly. He was forced to wade through the water just to get to the door. His wife, Jamima, was about eight months pregnant, and his son, Major, was only three years old. They needed their husband and father desperately. Water was flowing into the house, forcing Johnson to move Jamima and Major to a higher room. He scrambled to get them to safety and to salvage any personal items he could. The rising water had trapped them in their own home, and Johnson was no longer watching the track.

The track was already in need of attention. At 11:08 p.m., around the same time John Johnson had first left his house for work, a freight train left the town of Buford, roughly seven miles to the west of Thaxton. A conductor named Butler was in charge, and a cattle dealer named T. P. Ayers was aboard the train with his livestock. Ayers had boarded the train at Abingdon near the Virginia-Tennessee border around ten o'clock that morning and was resting in the caboose. About halfway between Buford and Thaxton, the clamor of rain pelting the train stirred Ayers from his sleep. He got to his feet and gazed out of the rear window. The lights mounted on each side of the caboose illuminated the heavy rainfall, and Ayers could see water running down both sides of the track.

Ayers continued to watch the elements through the windows of the caboose as the train passed the station at Thaxton. The freight train was about a mile beyond the station when the engineer slammed the engine into reverse, hurling Ayers to the front of the caboose. He gathered himself, fortunately not any worse for the wear, and stepped out of the caboose to investigate the situation. A telegraph pole had fallen across the track, and the engine had become entangled in the lines. Conductor Butler worked with his engineer to clear the wires and pulled them off trackside. Knowing that passenger train Number Two would be coming along behind them soon, Butler was especially concerned about the condition of the track. He instructed his flagman, John Read, to go back to the Thaxton station and alert the dispatcher about the flooding they had seen and about the issue with the telegraph pole. The conductor also instructed flagman Read to stop train Number Two, alert them to the situation, and warn them to look out for the freight train ahead in case it had to stop again. Butler knew that it was a dangerous night on the rails. He urged his flagman repeatedly to make sure he notified passenger train Number Two about the problems on the railroad. It was about 11:30 p.m., and John Read headed back for Thaxton.

One of the jobs of the flagman on a train was to be ready to go quickly on foot behind or ahead of a train in the event that some unforeseen issue caused it to stop. He would typically carry a red lantern, white lantern, fuses (flares with a spike that he could drive into a wooden railroad tie), and "torpedoes." A torpedo was a small explosive device that a flagman could attach to the rail. When the wheels of a train passed over a torpedo, it would create a very loud bang and alert the engineer that there was trouble ahead and to slow down. It was an extremely dark night with very little visibility. Read would need to use all options available to signal and stop the passenger train to avoid a potential collision.

Flagman Read headed back west toward the Thaxton station. The telegraph poles had fallen around milepost 233, and the station was about a mile away near milepost 234. As he made his way back, he noticed water coming over the track at some points, but thankfully, there were no washouts. He passed by the station and, after going a bit further, put down one of his torpedoes as a signal to passenger train Number Two when it came along. Read returned to the station, set his red lantern down by the track, and summoned the telegraph operator. He informed the operator about the flooding, the downed telegraph wires, and Conductor Butler's recommendation to slow down any trains as they moved through Thaxton. The rain was coming down hard, which persuaded Read to stay at the station for the time being. His freight train had continued east to Liberty, where they would wait for him to return and for passenger train Number Two to pass. The operator sent a telegram westward to the Blue Ridge station to alert the passenger train of the issues, which would give the men running the train plenty of advance warning.

It was close to midnight, and the passenger train was scheduled to leave the city of Roanoke and head east at any moment. Roanoke was twenty-three miles away from Thaxton, and steps had been taken to protect the train from the dangers that lurked to the east of Thaxton. Read's lantern and torpedoes were in position to slow the passenger train down before it arrived at the station, and the telegram to Blue Ridge would ensure that the men in charge of the train were aware of the issues. It all seemed like a solid plan, but no one realized that watchman Johnson was trapped in his house, Wolf Creek was swelling by the minute, and passenger train Number Two would never make it to John Read's signals.

2

DEPARTURE

"We will leave Roanoke on prompt time, which is something unusual for this train."[1]

— JAMES C. CASSELL, SUPERINTENDENT, NORFOLK &
WESTERN LYNCHBURG DIVISION, JULY 1, 1889

James Calder Cassell was destined to be a railroad man. He was born in Lancaster County, Pennsylvania, in 1856, and his father was one of the oldest employees of the great Pennsylvania Railroad Company. At the age of fourteen, the younger Cassell was already working as a telegraph operator for the "Pennsy." He later moved on to work as train master for the Shenandoah Valley Railroad, which would eventually lead him into Roanoke, Virginia, in the employ of the Norfolk & Western Railroad.

Late in the evening on July 1, 1889, Cassell and fellow N&W Superintendent Edmund L. DuBarry were on the platform of the Roanoke passenger station, about to board passenger train Number Two. Already in his mid-forties, Edmund was the senior member in the pairing, but the men shared a similar career path. Like Cassell, DuBarry was a veteran employee of the railroad industry, and he too began his career with the Pennsylvania Railroad. DuBarry was superintendent over Norfolk and Western's Eastern Division, which covered the road from Norfolk to Crewe, Virginia. Cassell's Lynchburg Division connected from Crewe and continued west to Roanoke.

Train Number Two was traveling on a road originally constructed as part of the Virginia & Tennessee Railroad. The V&T opened in 1857 and ran from Lynchburg, Virginia, to Bristol, Tennessee. In 1870, the Virginia legislature passed an act to consolidate that stretch, along with the Norfolk & Petersburg, Southside Railroad, and the Virginia & Kentucky railroad to create one new conglomerate. This experiment was called the Atlantic, Mississippi & Ohio Railroad, but it was short-lived. By 1879, the new company was already having difficulties paying its debt, and the United States Circuit Court in Norfolk ordered the AM&O to pay the debts or sell off the railroad. In 1881, Clarence H. Clark and his associates paid over $8.5 million for the Atlantic, Mississippi & Ohio, and it was organized into the Norfolk & Western Railroad Company. In 1889, Norfolk & Western's road extended from Norfolk, Virginia, in the east to the city of Bristol, which straddled the border of Tennessee and Virginia in the west.

Cassell and DuBarry were boarding the train at the western edge of Cassell's division. The route ran east out of Roanoke, and passengers could connect to other railroad lines along the way. Some could head north on the Virginia Midland Railroad at Lynchburg or continue toward Norfolk and the Atlantic Ocean.

Their starting point that night, Roanoke, was a community truly born of the railroads. It had previously been known as "Big Lick," named after a prominent salt lick in the area popular with local wildlife. Efforts to rename the town began once plans for making it a railroad hub were put into place. The Norfolk & Western's east–west road connected at Roanoke with the north–south line of the Shenandoah Valley Railroad, up to Hagerstown, Maryland. In 1882, the Indian name for the river that flowed through the area, "Roanoke," was officially adopted for the newly flourishing city.[2]

Beyond Roanoke, the train's route headed toward Thaxton and passed several smaller stations, which included Bonsack, Blue Ridge, and Buford. Blue Ridge was also called Blue Ridge Springs and was well known to many around the country. A resort was built at Blue Ridge around its famous mineral springs, and it was particularly popular with wealthy vacationers. They came to the resort for the fine cuisine and the mineral waters believed to have medicinal benefits. The owner of the resort, Philip Brown, actually bottled and delivered the water from the springs to area residents for twenty to thirty cents per gallon, depending on the size of the order. He marketed the water as the "Celebrated Dyspepsia Water."[3] The Blue Ridge mineral water was said to provide relief for a number of ailments, including upset stomachs, kidney disease, and nervous disorders. Until 1888 one of the main attractions at the resort was the four-story hotel that featured a covered walkway connected to the railroad depot. The walkway was a popular spot for locals to do some people watching as distinguished visitors made their way from the station to the hotel. Sadly, an act of arson destroyed the hotel in the spring of 1888. The fire burned with such intensity that even the crossties supporting the nearby tracks were scorched, causing train delays for several hours. The hotel was still not fully reconstructed by the summer of 1889.

The route continued east from Blue Ridge toward Buford, and it was a demanding stretch of rail. When the road was constructed through this area, a legion of workers had used everything from shovels to black powder to carve a path through vast quantities of limestone. Even with all that work to clear the way, a steep incline remained for trains to overcome as they passed through Blue Ridge and tackled the five-mile trip to Buford. Locals referred to Buford by many names including Bufords, Bufords Depot, and Bufordsville. All these names originated from Captain Paschal Buford, the man who had deeded the right of way to the railroad through his land. Buford had struck a deal with the railroad to give it a right of way if the train would make regular stops at a depot in the town. The deal was made, and the town benefited from the boost in commercial activity. Like Blue Ridge, Buford also became a vacation destination for those seeking to escape the sweltering heat and humidity of the summers further south. A new hotel called the Glendower House had just opened at Buford. The hotel was capable of accommodating one hundred guests seeking the clean air and vistas of the Blue Ridge Mountains. Once passenger train Number Two made its way past Buford, its journey would take it only another seven miles further east to Thaxton.

On July 1, 1889, the train was scheduled to leave Roanoke at its regular departure time, 11:56 p.m. It was not out of the ordinary for the train to be delayed, and Cassell quipped to DuBarry, "We will leave Roanoke on prompt time, which is something unusual for this train"[4] as they boarded.

Delays were certainly not due to the capabilities of the engine pulling train Number Two. Leading the charge on that night was engine Number 30. Baldwin Locomotive Works of Pennsylvania built the engine in 1887, and it was known as the fastest passenger engine in service. Baldwin had painted the engine as black as the very night they were traveling, with "Norfolk & Western" printed in gold letters down the side of its coal tender. Engine 30 featured four big driving wheels that were over five feet tall, which enabled it to gain the speed and power needed to pull through the steep mountain grades of southwestern Virginia with ease. There were four smaller wheels on a pilot truck at the front of the engine to guide it down the track. The engineer rode in a cab constructed of white pine situated above the driving wheels, and the eight-wheeled coal tender made of oak trailed closely behind.

Seven cars were along for the ride provided by engine Number 30. Directly behind the engine and its tender was postal car Number 280. This location was traditional for most postal cars, in part to offer additional protection for any passenger cars further back on the train in case of an accident. Railroad cars close to the engine were more likely to be impacted in an accident, and, unfortunately, this was confirmed by the statistics. Postal clerks working in the cars were killed in a larger percentage than railroad employees working in other sections of the train.[5]

Next in line was East Tennessee, Virginia, and Georgia (ETV&G) baggage and express car Number 57, followed by second-class coach Number 54, first-class coach Number 63, and three sleeping cars operated by the Pullman Palace Car Company. The ETV&G baggage car was likely connected at the beginning of the N&W line at Bristol, where the two railroads met. The first of the sleepers was called "Beverly," followed by "Toboco," and finally "Calmar" at the rear of the train. Calmar was one of the newer "state of the art" sleepers manufactured by the Pullman Company. Outfitted with furnishings made of mahogany, fine carpeting, and luxurious upholstery, the sleeper included twelve sections and a drawing room. Calmar was a "through car" from New Orleans, and its destination was Washington, DC. It was originally designed for the "Montezuma Special," a luxury train that connected the national capitals of Washington, DC, and Mexico City.[6]

The Calmar car seemed like a good place to get the journey started for superintendents Cassell and DuBarry. They sat down together in the vestibule and waited for some of the ladies to get settled in their berths, while Engineer Pat Donovan and Fireman James Bruce went to work in the cab of engine 30. In the sheer darkness of a cloudy nineteenth-century night, Donovan put his hand to the throttle, and N&W passenger train Number Two set out on its journey toward Thaxton. Over seventy passengers and crew members were aboard.

3

SOULS

"Auntie sends love and so do I and a thousand kisses for you."[1]

— PATTIE LOVE CARRINGTON, PASSENGER ON TRAIN NUMBER TWO IN A LETTER TO HER GRANDMOTHER, JANUARY 11, 1889

The job of a locomotive engineer was not for the faint of heart. There were so many dangers to watch for when riding the rails that an engineer's personal safety was least on his mind when he stepped into his cab. A casual observer might have looked at these hulking chariots of iron and smoke and expected them to be nearly invincible, but they could be as fragile as glass under certain conditions. The slightest variation in the symphony of moving parts that made up rail transportation could result in disaster. A broken or loose rail, livestock meandering onto the track, undermining of a bridge or fill by flash floods, or even a mischievous child placing a spike on the rail could easily cause a locomotive to derail. Collisions with other trains were also fairly common due to the crude traffic control systems of the era. Telegraph systems relayed special orders from station to station, and flagmen with lanterns standing by the track signaled trains of impending danger. These methods were less than reliable, and a derailment or collision typically resulted in significant damages to life and property. The engineer was often crushed by his own engine and scalded to death by the escaping steam from the shattered engine's boiler.

Much like soldiers, policemen, or firemen, the railroad engineer's job was dangerous, and that was part of the thrill of working it. The men who work those jobs are usually made up of a proper mix of heroic ingredients. Pat Donovan was a young Irishman about thirty years old, but his experience made him well aware of the risks of his chosen occupation. To become the engineer of a passenger train, he had to work his way up through the ranks. An engineer typically would have worked in the shops to ensure that he knew every detail of his engine and be prepared to resolve any mechanical issues that might arise as they traveled down the road. He would also work as a fireman to learn the operations and procedures involved inside the cab until he put down his coal scoop to take the engineer's position. Once a man was ready to pilot an engine, he would cut his teeth on a freight train for a while before ever stepping into a passenger train to carry the most precious cargo on the line, human life.

Engineer Donovan had men, women, and children on his train, but he also had a schedule to keep. Along with his fireman, James Bruce, he pushed the train east toward the village of Bonsack. Bruce was feeding the hungry engine its midnight snack of coal. He was keenly aware of the dangers involved with his own job and had actually just renewed his $1,000 accident insurance policy the week before. No man would ever want to use a policy like that, but life on the railroad tended to be a little bit shorter for those who made their living on the engine.

Behind the controlled chaos of the engine operation, postal clerks Lewis P. Summers and James J. Rose were hard at work in the postal car sorting mail and papers for delivery at locations along the route. Both men lived in Abingdon, Virginia, about twenty miles northeast of N&W's western terminus of Bristol on the Virginia-Tennessee border. Their job was to get the mail sorted for its proper destination, which included mail that had already been loaded before the train left Bristol and sacks picked up along the way.

Since speed and time were always critical, postal clerks performed a unique maneuver to deliver and pick up mail at smaller stations. When the train was not expected to make a stop at a station, mail that was to be picked up would be hung in a sack from the arm of a crane alongside the track. As the train approached, the mail clerk would extend a hooked catcher arm from the car that would snatch the bag from the crane. The clerk would simultaneously kick out a mailbag destined for that location and an employee at the station would retrieve it. If the clerk did not kick with enough force, the bag would fall under the train and the wheels would rip the mail to shreds. Clerks referred to this unfortunate situation as a "snowstorm."[2]

Lewis Summers was still adjusting to the challenges associated with working in a swaying, lurching, and crudely lit postal car. He had just been appointed to his position as postal clerk on the run from Bristol to Lynchburg two months earlier. Railroad postal clerk positions were political appointments, and Summers was a very active member of Abingdon's Republican Party. When Republican President Benjamin Harrison took office in 1889, the Democrats who were holding those positions under Grover Cleveland's administration quickly found themselves unemployed. Not surprisingly, they were replaced with clerks loyal to the Republican Party. Lewis was one of those new appointments, but his job was much more than just a cushy government position. Postal clerks needed strong mental ability to learn, retain, and process the ever-changing information necessary to ensure that mail would be delivered to the proper location. Summers had trained with the outgoing clerks for a few weeks before taking over the route in June, and the job required him to make three round trips from Bristol to Lynchburg every two weeks.

His colleague, James Rose, was not one of the politically appointed postal clerks. He was assisting Lewis in order to learn the route and hopefully open the door to future opportunities. Both men had been busy on the trip receiving, delivering, distributing, and tying up mail sacks. One would have to forgive James if he had sorted a letter or two incorrectly that night. His mind was on other things. Young James Rose was in love, and he was one day shy of marrying the woman he had pursued for quite some time, Lillian May Grubbs. James had proposed to her several times in the past only to be rebuffed. She felt that James was a bit too wild for her tastes. He assured her that he intended to "lead a better life" and that "if he were called to die he would be saved," which convinced Miss Grubbs to change her mind, and she finally accepted his request for her hand in marriage.[3] Everything including the minister was in place for the ceremony the next day. James Rose needed only to finish the evening run and return to Bristol, where his bride would be waiting at the altar.

Just behind the postal car and mixed in with piles of suitcases, bags, and trunks in the baggage car were Robert Ashmore, William Graw, and Captain William Henry Ford. Ford was a grey-bearded man in his fifties and had been working with the railroad in various capacities for twenty-five years. His role on that night was that of the baggage master, sometimes more affectionately referred to as the "baggage smasher." The reputation of baggage handlers on the railroad was so well known that some companies advertised their luggage as so sturdy that it was "baggage smasher proof."[4] Ford had started his destructive work of recording and not so carefully stacking the passengers' baggage about thirty minutes before the train left Roanoke. The stacking was generally designed to ensure orderly offloading at the right stops along the line.

One of Ford's companions in the baggage car was Robert Ashmore. Mr. Ashmore was not in the business of torturing innocent luggage but instead worked as an express messenger for the Southern Express Company. His duty was to manage and protect the packages that were being shipped in the baggage car by his company. Southern Express was a separate company from Norfolk & Western, and its primary function was the safe transport of packages that were not carried by the postal service. Some trains had entire cars dedicated solely to "express" packages, but in this case the general baggage and the express items were combined on the baggage car. Among the items under Ashmore's charge was a safe containing bank and treasury notes worth several thousand dollars, along with several watches and pieces of jewelry. Typically, if a train was going to be robbed, the express company's cargo was exactly what the robbers were after. The dangers of the job were not enough to scare off Ashmore's coworker, William Graw. Graw was training as an extra express messenger and hoped to earn his own position with Southern Express.

Following the baggage car was the second-class coach, also called the "smoker," which was designated as the car that passengers from other coaches could move to when they wished to light up. Sitting amidst the cloud of smoke was an isolated black man named Robert Davis. He had the unenviable distinction of being the only passenger listed as "colored" on the train that night. The civil war had ended twenty-four years earlier, but the nation was far from thinking about treating its citizens equally.

Davis was no stranger to the challenges that a black man faced traveling the rails in a still heavily segregated world. The route he was traveling on that night was a key component of the Confederate supply line during the war, and many of the men he traveled with were Confederate army veterans. It took a special kind of courage to willingly subject himself to the cultural trials of his time, and his courage may well have come from his faith. Davis was the presiding elder of the Danville District of the African Methodist Episcopal Church and had been in the ministry for twenty years. He lived on the Eastern Shore of Virginia in the town of Eastville, but his work with the church required him to travel all over the state, including Roanoke. Cave Spring A.M.E. church was the first black church in Roanoke County after the Civil War, and it was part of Davis's district.

Trailing behind the second-class coach in position, but not in status, was the first-class coach. After the train left Roanoke, there were at least sixteen passengers in first class. Their destinations were as varied as their ages and occupations. This particular train connected passengers with steamer ship lines at Baltimore, New York, and Boston, and many of the passengers were on their way to cross the Atlantic. While some of those on board might have been brimming with excitement over an ocean cruise in their future, others were just as excited to reach their own blissful destination, home.

Conductor Roland Johnson made his way through the first-class coach checking tickets and making conversation with passengers as he went. Roland was about thirty years old and had worked his way up from clerk to conductor. The growth of the railroad across the country provided many opportunities for advancement to men who were industrious enough to take advantage. Roland was one of those hardworking men.

As he walked through the coach, Roland stopped to look at the ticket of Harry B. Wheeler, a small man sporting a brown moustache, a straw hat, and a salt-and-pepper–colored suit. Wheeler carried an annual railroad pass which meant he was obviously a regular traveler. Roland noticed that Wheeler's pass listed him as a "line accountant."[5] The conductor mistakenly thought Wheeler worked on the telegraph lines and joked that they were lucky to have a telegraph lineman on board since the wires were working badly. Wheeler's job was actually to travel the rails as an auditor for the Union News Company out of Richmond, Virginia. He was responsible for checking on the company's newsstands at various railroad stations to ensure that they were in proper order. Harry had been promoted to the position just months earlier and had moved from Baltimore to Richmond with his wife and young son. Wheeler's work kept him riding the rails for two weeks out of every other month. He was finished with his most recent trip and headed for home.

Conductor Johnson stopped to talk to another Richmond businessman on the first-class coach, Mr. John I. Stevenson. Mr. Stevenson and his boss, John Bowers, had left Richmond the previous Wednesday. Bowers ran a business that dealt in stoves, plumbing, and gas fixtures among other things. Stevenson and Bowers were working on a large contract job that required them to travel to Big Stone Gap in the southwestern corner of Virginia. Before the two men left Richmond, a salesman approached them offering accident insurance policies. This investment was generally worthwhile given the dangerous nature of rail travel at the time. Bowers decided to take up the salesman on his offer, but Stevenson declined. After completing their business at Big Stone Gap, they returned to Bristol and parted ways. Bowers headed further into Tennessee to do some additional business, and Stevenson set out for home to see his wife, Ada. Conductor Johnson and John Stevenson were both Freemasons, and they exchanged Masonic cards before Johnson moved on.

Conductor Johnson also found Fred Dexter traveling on business in the first-class coach. Dexter was in his early twenties and worked as a traveling salesman selling shoes for Burley and Usher in Newburyport, Massachusetts. He had boarded the train at Roanoke along with about $350 worth of his shoe samples. He settled down for a nap after a busy workday.

Like Dexter, a grey-eyed young man with brown hair and moustache to match had boarded the first-class coach at Roanoke. The young man's name was Nathan Cohen, and Roanoke was his new home. When Cohen moved to Roanoke from Baltimore two years earlier, he found employment with the Philadelphia One Price Clothing House. Although he was only about twenty-five years old, Nathan had already experienced more of the world than some men twice his age. He emigrated from Germany to the United States in 1881 and became a citizen in 1888. Nathan's trip on the train was just the first leg of a long journey back to Bremen, Germany, for a visit with his parents.

John Kirkpatrick and his friend Frank Tanner boarded the first-class coach at Roanoke after a long, frustrating day. Kirkpatrick and Tanner were both from nearby Lynchburg, and their trip to Roanoke that day had not been a planned one. Kirkpatrick was about twenty-five years old and worked as cashier and bookkeeper for the Ivey & Kirkpatrick Insurance Agency in Lynchburg. He worked there with his brother, Benjamin Kirkpatrick. John was well thought of in his hometown, and many considered him one of the most intelligent and generous citizens in the city. A friend had exploited that generosity when he approached John to have him endorse some checks for him. Unfortunately, the checks bounced, and the banks processing them in Roanoke were not happy. Kirkpatrick had traveled to Roanoke to resolve the matter, and his friend Frank Tanner accompanied him. On top of all the troubles of the day, Tanner and Kirkpatrick had missed the train that they had intended to take back to Lynchburg. Their plan had been to catch the train that departed Roanoke at 7:00 p.m., but they missed that train and instead were forced to settle on the midnight departure of passenger train Number Two. If everything went according to schedule the rest of the night, they could still get back to Lynchburg about one thirty in the morning.

The reason for travel was much more cheerful for three friends from Cleveland, Tennessee, on their way to visit Europe and the Holy Land. One of the most anticipated sights they planned to see was the Paris Exposition, a World's Fair that featured the newly constructed Eiffel Tower as its entrance gate. Will Marshall, Will Steed, and John Hardwick had been planning their trip for many months. The trio had been traveling since eleven o'clock that morning, beginning on the ETV&G railroad. After supper at Bristol, they had continued their trip on Norfolk & Western's road.

A trip to Europe was no small expense, but these young men had put the resources together to go. John Hardwick was just in his early thirties, but he was already a successful businessman. He was a founding partner of the Cleveland Stove Works along with his brother Joseph and his father Christopher. One of Hardwick's traveling companions, Will Steed, was also in his early thirties and already a partner with his brother in the J. A. Steed & Brothers drug store in Cleveland. Twenty-one-year-old Will Marshall was the youngest member in the group. Marshall was the Cleveland City Recorder, secretary at his father's lumber company, and vestryman and treasurer at St. Luke's Episcopal Church in Cleveland. All three men were high-achieving, eligible bachelors and well respected in the Cleveland community. They sat facing one another with one of the seats turned so that two of them were riding backward. The third man had his window down, allowing the warm night air to swirl inside the car.

Elsewhere in the coach, a young man about thirty years old with an impressive dark-brown moustache had wedged himself into a seat with his knees pressed against the seat in front of him. This tall man was Fred Temple, a Chicagoan on his way to New York to accept a job as a civil engineer. Temple had been in Chattanooga visiting friends and his bride-to-be, Mary Sherwood. He drifted in and out of sleep, perhaps dreaming of the wonderful things that were still to come in his personal and professional life.

A young family of three also rode along on the first-class coach with the businessmen and vacationers. Charles Peyton, his wife Jessie, and their infant daughter Charlene had probably boarded at Radford, Virginia. Radford was the halfway point between Bristol and Lynchburg and about forty-five miles southwest of Roanoke. Charles had family in the northern Virginia area, but he worked in Radford as a stenographer for Norfolk & Western. He and Jessie were only in their mid-twenties and just beginning their lives together. They had been married for about three years, and little Charlene was less than a year old.

Charles was not the only Norfolk & Western employee taking a pleasure trip on the first-class coach that evening. Dennis Mallon was the janitor at the N&W general office building in Roanoke, Robert Goodfellow was a clerk in the General Manager's office, and James Lifsey worked as a dispatcher in Cassell's Lynchburg Division. Since Lifsey was a dispatcher on the route they were traveling, he was keenly aware of the path they were taking that night. If any trouble arose, he would be one of the first to recognize it.

As was the case with all of the passenger cars, there was quite a mixture of hometowns, occupations, destinations, and reasons for traveling among the people on the first-class coach. Regardless of the reasons that brought them together, their life stories would forever include at least one new shared chapter after the train reached Thaxton.

Riding in a sleeper car was an experience quite a bit more relaxed than the cramped quarters of the day coaches. The cost of having a place to lay your weary head was higher, but well worth it if you were on a long journey that required a nighttime ride on the train. Compared to standard coaches, Pullman sleeper cars were big, heavy, and luxurious. They were the brainchild of George Pullman, a native New Yorker. Pullman envisioned a new way to build railroad cars to accommodate overnight travel. His idea was to create an "upper berth" where bedding and materials could be stored during day travel and folded down at night to create a private sleeping area. The berths folded up into overhead compartments to make room for standard seating during the day.

This superb innovation solved the lack of privacy in the open bunks used previously by travelers who needed a place to rest at night.

Sleeper cars would typically be staffed by a conductor and at least one porter, and both would be employees of the Pullman Company. While the conductor may have "officially" been in charge of the sleeper, the porter carried more sway than one might imagine. A former conductor described the porter in a sleeper as the man "whose word is law and by whose frown or favor the passenger is either very comfortable or supremely unhappy."[6] Porters generally prepared the berths for sleeping and packed them up once passengers awakened. They also provided other services, such as serving beverages or giving shoe shines. These men often had to deal with unruly passengers who tipped very little, but it was still better employment than could be found elsewhere. Despite the quirks of human nature, Pullman's sleeper car concept was a major innovation in travel. It also created an opportunity for women and children to take advantage of night trains that were previously the domain of men only.

There were several ladies taking advantage of Pullman's sleeper car service on the N&W that evening. Mrs. Roberta B. Powell and her daughter Inez Sparkman were from Texas, and they were riding on "Beverly," the first sleeper car behind the day coaches. Inez was only about eighteen years old. She and her mother had been vacationing for a few weeks near Roanoke with her friend Janie Caven.

Janie was a Texas gal as well. She was seventeen and attending school at Montgomery Female College in Christiansburg, Virginia, about thirty-five miles southwest of Roanoke. Her father, William Caven, was a prominent man back in Texas, where he was a successful real estate investor, farmer, and businessman. Mr. Caven was also a Confederate veteran and had served in the Texas state legislature. Janie, Roberta, and Inez were headed to Fredericksburg, Virginia.

The Beverly sleeper was not exclusively occupied by the ladies. Mr. Burton Marye was a civil engineer from Richmond in his mid-twenties. He was also the son of the state auditor in Virginia, Morton Marye. On a dark, rainy night, Burton may have been a little more nervous because he was quite familiar with the problems that could arise with bridges and culverts. Marye had spent some time working with the engineering corps of the Baltimore and Ohio railroad in Pennsylvania.

Bound for Norfolk, passengers going the full distance from Roanoke on the Beverly sleeper had 250 miles and thirteen more hours on the rails ahead of them. Some passengers, like John Rowntree, were already asleep in their berths before the train pulled away from Roanoke. Rowntree was a thirty-five-year-old hardware buyer for wholesaler George Brown in Knoxville, Tennessee. He was on his way to New York to do some purchasing for his company.

In addition to Rowntree and Marye, at least six other men would share the Beverly sleeper car with the three women from Texas. When the time came for action, not all of them would conduct themselves like gentlemen that night, but one of the ladies would outshine them all.

Washington, DC, was the destination for a sleeper car named "Toboco" that was coupled behind Beverly. At least ten passengers were aboard Toboco, and they were a mixture of young and old. John Irby Hurt and his sister Rosa Lee of Abingdon, Virginia, were representatives of the younger generation. Both were in their early twenties, and big brother John was working toward a career as an attorney. John Irby and Rosa Lee were still single, with their lives very much ahead of them. They were one of two sets of siblings on the train that night.

The youngest member of the Toboco family was a blue-eyed, blonde-haired, nine-year-old girl named Pattie Carrington. She was traveling with her "Aunt Kate" all the way from Texas to Virginia. Despite the long trip, little Pattie had been cheerfully playing around in the sleeper all afternoon to the delight of some of the passengers. The lengthy trip was really no big challenge for young Pattie because she had traveled back and forth from Texas to Virginia many times. Her Aunt Kate was actually her great aunt, Catherine Lightfoot Carrington Thompson. Aunt Kate was from Virginia, and her nephew was Pattie's father, a Confederate veteran and a successful attorney in Houston named William Allen Carrington. He was known by his middle name, Allen, and he had suffered unimaginable sorrow in his lifetime. His first wife Martha, Pattie's mother, died due to complications from Pattie's birth. Aunt Kate stepped in and had essentially adopted Pattie from infancy. In the years following Pattie's birth, her father lost a second wife, and death took Pattie's older brother "Fonny" at the age of eleven. Pattie's father summarized his life in a letter to his sister in 1885 as "having sounded to its depths all the Hell that this world can contain or inflict within sufferable bounds."[7] In spite of all the despair in his life, he had not lost all hope. Just a few weeks before Pattie and Aunt Kate boarded train Number Two, Pattie's father married for a third time. He and his new bride had set off on a much-needed trip to Europe. It seemed possible that Allen Carrington might capture some joy in his world after all.

One of the passengers who enjoyed watching Pattie play in the sleeper was Edith Hardester. She and her friend Annie Fishpaugh boarded the train at Knoxville around one thirty that afternoon. Edith and Annie were both young ladies from Baltimore and worked together as milliners designing and making hats for the Young & Williams Company in Knoxville, Tennessee. They were on their way to visit relatives in Baltimore and had decided to share a berth in the sleeper for the trip.

Strangely enough, if you desired a fashionable hat, Toboco was the place to be that night. Edith was accompanied by another friend named Irene Jackson. Irene was the head milliner at a different firm in Knoxville, Lon Mitchell's "Dry Goods, Boot and Shoe and Millinery Emporium."[8] She oversaw the design and creation of ladies' hats for the firm. Like Edith and Annie, Irene was single, but she didn't plan to stay that way for long. Miss Jackson was on her way to her hometown in Cambridge, Maryland, to become Mrs. Wilkes. The lucky groom was Mr. Eugene Wilkes from Laurens, South Carolina. As long as there were no setbacks, Eugene and Irene would be married on July 17.

In the berth directly across from Annie and Edith was a man who was a veteran traveler, Alpheus Waters Wilson. After a trip to preach at the North Georgia Ladies' Missionary Conference in Dalton, Georgia, Wilson was returning to his home in Baltimore. He traveled so much that one of the railroad companies should have named a depot after the man in some town along the line. He was in his mid-fifties and had been a bishop for the Methodist Episcopal Church, South for seven years.

The goals and demands of his position in the church required him to travel around the world. Bishop Wilson had a passion for missionary work and spreading the message of Jesus Christ to the world. He spent much of 1888 on a missionary trip to Japan, China, India, and other eastern nations to help establish churches and schools. His travel schedule in America and overseas meant that his was a familiar face at many railway stations across the country.

Alpheus was not just a man who simply traveled about handling the administrative duties of the church, however. Bishop Wilson had preaching in his blood. He was the son of a preacher, and his wife, Susan, was a preacher's daughter. Alpheus Waters Wilson was considered one of the most extraordinary preachers of his time. According to some, his skill as a preacher might have been powerful enough to prevent the Civil War.

Wilson was heavily involved with the Methodist church when, like the rest of the nation, the church split internally during the Civil War. He remained loyal to the Baltimore group that included churches from Virginia and below the Mason-Dixon Line, and he would not succumb to the political pressures of the time. The internal strife among the Methodists led to the occasional appearance of Union officers at Bishop Wilson's church to intimidate him. One evening Union general John Dix visited, and ministers loyal to the northern faction of the church urged Dix to arrest the Bishop. After hearing a sermon from Bishop Wilson, General Dix would not comply and told the ministers, "if his ministerial accusers had preached the gospel as he preached it, we should not have had this war."[9]

Wilson's passion and effectiveness as a preacher were known around the world. The pastor of a church in Shanghai once said that "Bishop Wilson was the greatest preacher he had ever heard, and he had heard many."[10] Even Josephus Daniels, a man who would later become United States Secretary of the Navy, said that he heard the bishop "preach the greatest sermon to which I have listened. After the passing of a fifth of a century that sermon remains with me still. It has helped me all the succeeding years."[11]

On the train that night, Bishop Wilson carried his beloved New

Testament written in the original Greek language. It was said that Wilson would read it through every few weeks, and he took it with him on all his travels whether at home, abroad, on land, or at sea. Sleep would eventually trump his study of the Scriptures sometime before the train arrived at Thaxton.

The last car to disappear into the gloomy night as passenger train Number Two left Roanoke was the Pullman sleeper known as "Calmar." Calmar carried more passengers on the trip than any of the three sleepers, at least thirteen, along with two porters named Banks and Gambler. The passengers on Calmar hailed from many different locations, including New Orleans, San Antonio, Iowa, Philadelphia, and Tennessee.

A trio of young friends from wealthy families traveled together on Calmar. The youngest of the group was sixteen-year-old Reuben S. Payne Jr. who was traveling along with his sister Pauline and their friend Florence Vanuxem. Reuben and Pauline's father, Reuben Sr., was the president of the East Tennessee National Bank and a former mayor of Knoxville. The siblings were on their way to New York and Florence was returning to her home in Philadelphia. Her father, Frederick, had moved the family from Knoxville to Philadelphia to make his living with the Mutual Life Insurance Company.

Another Philadelphia native on board Calmar was young George Masters. Like Florence and the Paynes, George was the son of a wealthy father, David Masters, a partner in a clothing firm in Philadelphia. George was prepared to make his own way in the world as a civil engineer and had just graduated from college a month earlier.

Joining the young well-to-dos on Calmar were two off-duty railroad employees taking a trip on the train without the pressures of work. Charles Montague was normally a Pullman conductor, but he was on sick leave. For now, he could leave the joyful task of finding a temperature in the car that was just right for every single passenger to someone else. J. H. Elam was the other railroad employee taking a leisure trip on Calmar, but he was probably keeping his position as a "baggage smasher" for Norfolk & Western quiet. Elam not want to draw the ire of any passengers who might have experienced his handiwork on previous trips. On that night, he was on a furlough, so he left the destruction of pristine suitcases to the men who were on the clock.

The midnight leg for passenger train Number Two was officially underway as the lights of Calmar faded into the darkness beyond Roanoke. The crew was busy at work performing the functions they were paid to do and had done so many times before. The passengers in the sleepers were settled into their berths and drifting off to sleep or had already achieved it. Idle conversations may have taken place in the coaches, and some may have been thinking some happy thought about their destination ahead. Others may have just gazed out at the darkness of the valley surrounded by the Blue Ridge Mountains. There was astounding beauty to be seen there during the light of day, but in the middle of the night, there could be something ominous to it. The coal-black shadows of the mountains towered like walls of a fortress against the cloudy night sky and the mind might wonder what dangers lurked hidden in those shadows. The train seemed to provide some security against whatever imagined agents of terror lurked in the darkness. The sounds of an engine churning, wheels making a familiar squealing along the track, and wooden train cars swaying back and forth represented the sounds of humanity passing through the night. They were six miles from a small station at Bonsack and completely unaware of the calamity that waited for them ahead.

4

FOREWARNING

"I am of the opinion that the train was within half a mile of the bank when it washed out."[1]

— CAPT. W. H. STANLEY, NORFOLK & WESTERN
SUPERVISOR, LYNCHBURG DIVISION

Superintendents Cassell and DuBarry were sitting in the Calmar sleeper and had not been chatting long when they felt Engineer Donovan slow the train to a complete stop. They had arrived at the depot in the small village of Bonsack. The train was right on schedule. It was beginning to look like it might be a smooth trip after an on-time departure and an uneventful first few miles into Bonsack. The engine powered back up and pulled out toward Blue Ridge, about five miles away.

Not long after leaving Bonsack the train was moving slowly, and from time to time it would stop and start right back up again. Something odd was going on, and Cassell and DuBarry decided to investigate. DuBarry grabbed a lantern, and the men made their way through the sleeper cars and on to the platform of the first-class coach. They found small piles of dirt and debris washed down on the rails from rains that had passed through earlier that night. The rain had subsided, and trackmen quickly cleared the debris ahead of the engine so it could get moving. While DuBarry remained on the first-class platform to keep an eye on the track, Cassell decided to cross over to the second-class coach and get an update from Conductor Johnson. The train continued slowly while Engineer Donovan kept a sharp eye out for obstructions and potential washouts on the track.

The slower speed and periodic stops caused the train to arrive at Blue Ridge about two minutes late. While two minutes may not seem like much, the lost time could add up quickly if they continued to run into issues along the route. When the train stopped at Blue Ridge, Baggage Master William Ford was still at work sorting the passengers' luggage. He could hear rain pelting the tin roof of the baggage car, and he peered through the door to get a look at the weather. Water was covering the tracks, and Conductor Johnson was nearby, standing in the rain with an umbrella. It sounded like Johnson was speaking with the Blue Ridge telegraph operator.

"What is the matter?" Johnson cried out to the operator.[2]

"The track is flooded to Liberty, run with great care, there is great danger!" came the reply.[3] The telegraph operator was reading a telegram sent from Thaxton. It was the message sent less than an hour earlier that night by Conductor Butler, after his freight train had run into a fallen telegraph pole just beyond the Thaxton station.

Inside the mail car, Postal Clerk Lewis Summers overheard the distressing conversation between Johnson and the Blue Ridge telegraph operator. Summers surveyed the condition of the track from the door of his mail car. Water flowed over the rails, and he could barely make out the track as the water rippled over it. Several men scurried about with lanterns to inspect the condition of the road. As Summers returned to his work, he mentioned what he had overheard to his assistant, James Rose. Summers was not overly concerned about the situation, but Rose looked a bit nervous. James decided he would stay extra alert as they pushed forward. If the train did run into trouble, it was not uncommon for those who had enough warning to jump from the train at the last moment. A desperation jump in the midst of a wreck could mean going home with just a broken leg instead of inside a wooden box. Rose was taking no chances.

Conductor Johnson stepped out of the rain and onto the second-class coach, where he met with Superintendent Cassell. The telegram reported flooding east of Thaxton, and communication was down at least from Thaxton east to the town of Liberty, about five miles from Thaxton. Since that stretch of rail now had no communication, Cassell gave instructions to Johnson to stop the train at the next depot, which was Buford. Once he arrived there, Cassell hoped to determine the location of any trains that were east of Thaxton and make sure they managed the situation properly to avoid collisions.

With this plan in mind, they pressed on for Buford, about five miles east of Blue Ridge. Cassell headed back to the first-class coach to discuss the situation with DuBarry while the engine struggled to get going. The steep incline through the mountains was a challenge, especially when the train started from a complete stop. It was the first battle that engine Number 30 waged against the elements that night. In this case, man-made horsepower was up to the task and conquered the challenge presented by the mountain. They were further behind schedule, however, because the train pulled away six minutes late from the Blue Ridge station.

They had not gone far before Mother Nature once again taunted the train. About halfway between Blue Ridge and Buford, a watchman stood alongside the tracks, signaling with his red lantern for the train to stop. They were near a small community known as Ironville, and once again there was water running over the track. It was half past midnight, and it was becoming clear that the weather was determined to harass train Number Two for a while longer. Cassell got off the train to inspect the latest issue. He noticed that the water was up over the wooden railroad ties but not over the track itself. Since the area was part of his division, Cassell was familiar with it, and he had seen the water higher at Ironville on several occasions. He decided that it was safe for the train to proceed cautiously through the high water. They crept along as if they were working through an invisible traffic jam until the entire train had cleared the flooded area.

Unfortunately, they were still not ready for "full steam ahead" to Buford. Cassell knew that there was a road crossing just ahead where the section crew frequently battled with dirt and rocks washing down across the tracks. The crossing had been a nuisance recently thanks to the rainy spring and summer. Just to be on the safe side, Cassell asked Conductor Roland Johnson to step off and walk in front of the train with a lantern to examine the crossing. Roland tried his hand as a track inspector and found no issues. They quickly resumed the trip toward Buford.

After eavesdropping on the telegram conversation at Blue Ridge, Postal Clerk Lewis Summers had decided to sit by the door to his car until they cleared the heavy incline. Once he was comfortable that the train had made it through the trouble spots, he made one final check to ensure no letters were left to handle. His work for the night was done. He put away his supplies and tied up the remaining sacks. Summers wanted there to be no delay when the time came to board his return train home at Lynchburg later that night. Once he had everything in order, it was time for a quick nap. It was a little known fact, but sacks of United States mail made a mighty fine mattress when stacked just right on a table. Summers had already learned this trick, and blissful sleep quickly arrived for him near the back of the postal car on his own handmade bed stuffed with fluffy correspondence.

He was already fast asleep when the train pulled into Buford about twenty-seven minutes behind schedule. Any passengers still awake were likely frustrated by the slow pace and frequent stops. It had taken nearly twenty minutes just to cover the five miles from Blue Ridge to Buford. One of those restless passengers was Frank Tanner, and he was ready for a smoke. Tanner nudged his friend John Kirkpatrick and asked if he would like to accompany him to the second-class car to relieve some stress through the time-honored tradition of burning tobacco rolled in paper. Kirkpatrick was mentally and physically tired from his long day of wrestling with the bad check issue in Roanoke. He just wanted to go back to sleep. He handed Tanner a cigarette, and unknowingly the two friends spoke to one another for the last time.

Superintendent Cassell walked into the telegraph office at Buford to find out where the trains were located east of the town of Liberty. With this information in hand, he could give instructions to any westbound trains once he arrived at Liberty. If they were not careful, the telegraph outage between Thaxton and Liberty might lead to a collision with an oncoming train. After getting the information he needed, Cassell left the telegraph office and headed for engine Number 30. He planned to climb aboard with engineer Donovan and Fireman James Bruce for the ride from Buford to Thaxton. They needed to make sure that they took every precaution along the six or so miles between the two stations. It was not raining heavily in Buford, but their path to Thaxton included several creek and tributary crossings. It was a dangerous night, and the water crossings ahead might be hazardous. They could use as many watchful eyes as they could get.

Just as Cassell made his way in the direction of the engine, he saw Alvin James walking toward him. Alvin was the Road Foreman of Engines in Cassell's division, which meant he was a supervisor over all of the engineers. Norfolk & Western had promoted Alvin to the position just seven months earlier. He was traveling to Burkeville, Virginia, as a passenger that night. Perhaps as a way to impress his division superintendent, or just because he was the type of man who would step in when he saw a need, Alvin volunteered to ride on the engine instead of Cassell. He told Cassell that he would direct Donovan to run the engine carefully, and, in particular, he would have him run slowly over bridge number 125 at Goose Creek. It was located about halfway between Buford and Thaxton and was one of the largest of the creeks they had to cross. If there were going to be a danger spot, Alvin thought it would be Goose Creek. Just to be safe, he told Cassell he would have the engine run carefully beyond Goose Creek and all the way to Thaxton.

Cassell took Alvin up on his offer. He asked him to stop again at Thaxton so that he could issue further train movement orders from there. Alvin then headed for the engine, and James Cassell moved to the baggage car. He felt that a seat by the side door of the baggage car would be the best place for him to keep a close eye on the track conditions. Passenger train Number Two pulled away from Buford at one o'clock in the morning on July 2, and James Cassell and Alvin James had just exchanged fates.

5

DISASTER

"A small culvert may give way in the night without a moment's notice."[1]

— CHARLES W. FOLSOM, ENGINEER, JOURNAL OF THE
ASSOCIATION OF ENGINEERING SOCIETIES, VOL.5, 1886

Passenger train Number Two was thirty-two minutes behind schedule after leaving Buford. Potential danger was ahead, but the crew was vigilantly watching for trouble. While his fellow superintendent monitored things from the baggage car, Edmund DuBarry felt comfortable enough with the situation to head back to his sleeper and lie down. The crew appeared to have everything under control, and they were still a good distance from DuBarry's Eastern Division. He walked back to the Beverly sleeper, the first in line behind the day coaches, and settled into his berth.

About two miles to the east of Buford, the train approached the area that had concerned the railroad men the most that night. As the crew members nervously watched for signs of trouble, Engineer Patrick Donovan eased the train slowly across the one-hundred-foot-long bridge over Goose Creek. The big creek waited like an assassin in the darkness below, but the bridge was secure. Donovan kept the throttle low as they continued another mile to Price's Cut.

A cut was a section of rail where dirt was removed to allow the tracks to go straight through a hill rather than going around or over it. Price's Cut was about 150 yards long, and it was a known trouble spot for Norfolk & Western. Water would flow into the cut and leave driftwood on the track, especially when the storms were as fierce as they were that night. Engineer Donovan cautiously brought the train into the cut. The sound of steady water flowing filled the air. From his new position in the baggage car, Superintendent James Cassell could see water running briskly in the ditches alongside the tracks, but the rails were clear. Rain had obviously been heavy earlier that night to fill those ditches, but there was no rain to speak of at the moment.

As the train exited Price's Cut, Engineer Donovan still had the engine running at a cautious speed. He spotted one of the local track watchmen, a black man named Stephen Hurt, and called out to him. Hurt stood near a culvert, and Donovan wanted to know how high the water was at that point. Hurt replied that the water had been "mighty high"[2] earlier, but it had subsided. In fact, the water had been so high in Price's Cut around midnight that it had nearly drowned him. A powerful storm had swept through and produced flash flooding that caused Hurt to find himself hip deep before he knew what had hit him. If not for the help of a nearby man named Maxey, Hurt may well have drowned. It was the hardest rain the watchman had ever seen, and at one point, he was so terrified that he simply called out to the Lord to spare him.

All that water had cleared out, however, and it was no longer up over the rails. Despite the improved conditions, Donovan brought the train along so slowly that he was able to carry on his conversation with Hurt without even stopping.

Only a few small creeks and fills remained between Price's Cut and Thaxton, including the embankment called Newman's Fill over Wolf Creek. There was little reason for the men operating passenger train Number Two to worry about that section of track. In fact, Watchman Hurt considered it "as good a piece of track as we had on the division."[3]

As the engine and each of the seven cars rolled by, Hurt had no idea that he was likely the last man to see passenger train Number Two that night.

John Read, the flagman from the freight train that had tangled with the telegraph wires earlier, was still waiting at the Thaxton station. The hard rain around the station had subsided sometime shortly after 1:00 a.m. Read was just waiting for passenger train Number Two to arrive. His plan was to go out and signal the train to stop and let the crew know about the situation ahead. After a wait of nearly two hours, Read finally heard train Number Two approaching in the distance. He grabbed his red and white lamps and began walking down the track, toward the advancing train.

It was about 1:25 a.m. on July 2, and engine Number 30 approached Newman's Fill over the little stone culvert at Wolf Creek. Since they had moved so cautiously, it had taken the passenger train twenty-five minutes just to cover six and a half miles from Buford to Wolf Creek. Engineer Donovan had picked up the pace a little after they had navigated safely through the more dangerous areas of the line, but they were nearly an hour behind schedule.

Conductor Roland Johnson was in the second-class coach. He reached up to pull the whistle cord to alert Donovan to stop at Thaxton.

As he walked down the track, Flagman John Read heard the wail of the train's whistle blow. He waited to hear the sound of the signal torpedo he had placed on the track exploding next. It would never come.

Engineer Donovan guided his engine onto the fill at about twenty miles per hour. The massive amount of rain that had fallen that night was too much for the culvert beneath the fill to handle. In just a couple of hours, water had risen from a few inches deep to a depth of twenty-three feet on one side of Newman's Fill. The pressure of the water had wreaked havoc on the earthen fill, and the weight of the train was the tipping point. Donovan's engine had nearly made it across when the ground beneath it gave way. They were only a half mile from making their next stop, but there would be no arrival at Thaxton station on this night.

An enormous hole opened up under the train as the surge of water carried away clumps of saturated earth and the track lost its foundation. Horsepower could not save them, and the engine slid backward into the newly created abyss underneath it. The coal tender behind the engine folded over the top of its boiler like the jaws of an alligator clamping shut on its prey. The postal car and the baggage car snapped free from their couplings, and the baggage car telescoped its way into the postal car. Both cars violently came to rest on the left side of the once mighty engine.

The unsuspecting riders in the first and second-class coaches were next to take the awful plunge. The first-class coach split in half, just in front of civil engineer Fred Temple's seat. The force of the wreck squeezed his long frame between the seats and crushed his back as if he were inside the jaws of a vise. The car turned on its side, hurling Temple to the opposite side like a rag doll. A roaring mass of water, iron, and timber piled itself on Temple and twenty of his fellow passengers riding in the front coaches.

Edmund DuBarry had not yet drifted off to sleep in the Beverly sleeper just behind the coaches when he experienced the unforgettable feeling of train wheels slamming against railroad ties. He attempted to get up, but it was too late. The Beverly sleeper crashed and turned sideways but had not fallen all the way to the bottom of the chasm.

The weight of the cars falling into the pit caused the Toboco sleeper behind Beverly to break in half. The sleeper ripped into two pieces directly at the feet of Irene Jackson, one of the three young ladies who worked as a fashionable hat designer. She reached out for her friend, Edith Hardester, but could not see her in the blackness. She pulled herself back into the rear portion of Toboco and for the time being avoided a fall into the pit. Her two friends were not as lucky. Edith Hardester and Annie Fishpaugh, along with Bishop Alpheus Wilson, were all in the front half of Toboco. They were awakened from their slumber to the horrific sensation of falling into darkness.

The passengers and crew riding on the train's rear sleeper, Calmar, had absolutely no idea of the catastrophe that had just occurred ahead of them. The momentum of the wreck was not enough to carry Calmar off the rails and into the collapsed fill. All they felt was a sudden jerk, like air brakes applied to stop the train.

Samuel Boyd was sitting on a stool on the back platform of Calmar. He was the flagman and rear brakeman, which meant that he would manually apply brakes on the cars from the rear of the train forward if emergency stopping power were needed. While passengers and crew were being tossed about and buried in wreckage in the rest of the train, Boyd was not even knocked off his stool.

J. H. Elam, the furloughed N&W "baggage smasher," was also on Calmar. When the train came to an abrupt stop, Elam was walking toward the Toboco sleeper. The force of the impact to Calmar was so subtle that Elam was not even thrown from his feet. He stepped out on the front of Calmar to find out why they had stopped. He found the rear end of Toboco jammed onto the front platform of Calmar. He had almost no visibility, but he could hear screaming in the darkness. They were wrecked.

He leapt from the platform and ran to the back of Calmar to alert flagman Boyd. As Elam approached, Boyd was already on the ground with his signals. "We are wrecked, go back with the flag," he told Boyd.[4] Boyd took off in a full sprint westward toward Buford to stop any trains that might be approaching behind them. Neither man fully understood the magnitude of what had just occurred.

The chaos of the cars plummeting into the pit on top of one another carved a permanent signature into the senses of those who lived through it. The clang of iron slamming into iron, the sinister hissing of escaping steam from the boiler, the sounds of splintering wood, and the screams of terror and pain tortured their ears. The smells of muddy water and freshly overturned soil became frightful scents that would always bring back haunting memories of that night to the survivors. At first, the terrible blackness of the night protected their eyes from the images of death and destruction that surrounded them. Light would come soon enough, but the source of that light would only magnify the horror of the night to an unimaginable level.

Despite the full assault on the senses brought on by the crash, silence and darkness filled the moments just after the wreck had settled. Bishop

Alpheus Wilson found himself planted in the mud about four feet deep. The darkness was impenetrable to his eyes. A dreadful thought entered his mind that somehow he was the only survivor because he could hear and see nothing.

He was not alone. Annie Fishpaugh could see nothing and felt only the wet earth surrounding her. Suddenly, she heard the horrifying scream of her friend Irene Jackson fill the air. Irene fainted almost immediately, and her screams turned to silence. As passengers and crew members scattered about the wreckage began to get some sense of what had just happened, the cries for help and moans of agony began to spread. For those who had survived the collapse of the train into the washout, the fight for life had only just begun.

Fred Temple peered into the darkness from beneath the debris of the first-class coach. A bolt of lightning streaked across the sky and provided a fleeting glimpse of a possible escape route. He dragged himself to his hands and knees and crawled through an opening created when the coach split in half. Temple struggled to move with the intense pain in his back, but he managed to crawl through the wreckage into what was left of the Toboco sleeper. His strength could carry him no further. He collapsed into an empty seat.

Postal Clerk Lewis Summers awoke from his slumber to find himself in the darkness on his hands and knees in about one foot of water. His makeshift bed of postal sacks had been replaced with a pile of debris surrounding him. Just above his head he could barely make out an opening in the rubble. He managed to lift himself up and through the opening to escape the ruins of the postal car.

When the train went down, Superintendent Cassell had been sitting by the open door of the baggage car. The crash had thrown him to the opposite side of the car instead of pitching him out of the open door. He found himself mixed in amidst the wreckage of the demolished car, with a broken arm and terrible cuts and bruises. He called out for help.

Someone in the darkness replied, "Come this way!" It was Lewis Summers.[5]

"How, how? How did you get out?" Cassell cried.[6] The postal car and baggage car had collided and twisted themselves together, and Cassell was not far from the opening that Summers had crawled through. Summers guided Cassell to his escape route and assisted him up to the top of the debris pile. Cassell leaped from the pile and into the rushing waters of Wolf Creek.

Another cry for help rang out nearby. Summers groped his way through the darkness toward the voice of the baggage master, William Ford. Ford had tried to free himself, but something had him trapped. Summers needed to determine what was holding Ford captive, so he asked the baggage master for a match. There were matches in Ford's vest pocket, and Summers fished them out. Getting the matches was the easy part but finding something to strike a match on was no simple task. He fumbled around, looking for a dry enough piece of wood for the job. He found what he thought was a worthy candidate, slid the match across it and hoped for the best. The match flared up, briefly sweeping aside the curtain of blackness. Summers could see a heavy timber crushing down on the small of Ford's back. He lifted the timber with all the strength he could muster while Ford fought through the pain to slowly drag himself free. Ford thought his back might be broken.

"Pull me out!" he pleaded to Summers.[7] The postal clerk eased the injured baggage master out and began to pull him clear of the wrecked car.

Conductor Johnson had somehow survived the violent crash in the second-class coach. Like Ford, he was pinned down at first, but he was able to free himself. He dropped down through an opening in the bottom of the coach and into Wolf Creek. The current swept him helplessly away from the wreckage.

Edmund DuBarry would have liked to take his chances with the creek. The crash of his sleeper, Beverly, had also left him trapped. He was lying sideways, his head near a small hole in the side of the car about eight inches wide. DuBarry could feel it with his hand, and the night air breezed in through the hole. At once he heard the terrifying sound of escaping steam from the engine. The sound provided little warning for DuBarry before he began to feel the heat as it swirled in through the opening. He quickly pulled up his blanket to shield his face and hands. He held his breath and hoped that the blanket would be enough to defend against the assault on his flesh and airway by the blistering hot vapor. The steam rolled into the car and surrounded him, agonizingly scalding his left shoulder. The pain was terrible, but the blanket had saved him from a far worse fate for the moment. He exhaled a sigh of relief.

His victory celebration was short, and his ride was not finished. The Beverly sleeper was perched on the side of the washout and had not fallen to the bottom of the pit. Nearly ten minutes had passed since the wreck, and the fill beneath the heavy sleeper had continued to disintegrate. The weight of Beverly became too much for the embankment to hold, and the sleeper slammed down onto the first-class coach at the bottom of the washout. Beverly's fall pulverized the smaller coach, and anyone who had not already escaped was crushed alive. The fates of the three young men from Cleveland on their way to Europe, the Peyton family, and the others who purchased tickets for the awful ride in first class were not yet known.

Lewis Summers had just helped William Ford to his feet when they heard the alarming sound of Beverly's crash. Summers assisted Ford as they scrambled together to take shelter from whatever might be coming their way, but they soon realized that they were not in the path of the falling sleeper car. Summers helped Ford get to a safe spot away from the wreckage and then returned with the matches with the hope of finding anyone else who needed help. The matches provided little benefit in the unmerciful darkness. He called out, hoping to hear a reply from his assistant, James Rose, but no answer came.

Summers was frantically searching the wrecked car when suddenly he heard something from beneath the debris. It sounded like someone moving, perhaps breathing, but unable to talk. Could it possibly be James Rose? He blindly floundered his way through the shadows toward the source of the noise. His hope that the sounds were coming from his young helper was quickly dashed. The sounds were not those of a man, but instead Summers had found poultry. Chickens and ducks were scattered about the wreckage and must have been in baskets for transport in the baggage car. The newly liberated fowl had given Summers a false hope that he had found Rose. After his fruitless search, Summers climbed out of the wreckage and sat down on the eastern slope of the washout to catch his breath.

After his sleeper had fallen a second time, Superintendent Edmund DuBarry was pinned down tighter than he had been when the train first wrecked into the washout. The hole above him had opened wider, and his head was free from the car along with one hand. He could not move. DuBarry's bedding was soaked in the superheated steam that billowed from the engine. He was in excruciating pain, and the debris pressed his back hard into the scalding bedclothes. Then he heard the voice of someone crying out.

"Where are we? Where are we? My God! We will be all burnt up. I will be drowned. The water is below me."[8]

Despite his intense pain, DuBarry called out to the man and asked for his name.

"Who are you?" the man replied.[9]

DuBarry told the man his name and let him know that he was in the Beverly sleeper. The mystery man finally got his wits about him and told DuBarry his name. It was Burton Marye, the civil engineer from Richmond and the son of Virginia's state auditor. He had been asleep in the lower berth near the front of Beverly when the train crashed, and the upper berth had fallen and blocked his path to the aisle. The window next to his berth was his only possible escape route. Marye was not hurt, but he was gripped by the fear of drowning in the water below the car.

A few minutes of silence passed while he contemplated his situation, until the voice of someone outside the train called out to Marye. DuBarry could make out that the person was persuading Marye to drop out of the car into the water and swim to shore. Marye got his courage up, forced his way through the window in his berth, and dropped safely from the car. He had injured his wrist during his escape, but he was free.

"I am in the mud, but all right," Marye called out first to the unknown voice.[10] He then turned his attention back to DuBarry. "I will come back and help you as soon as I can get some men," he shouted to DuBarry.[11]

There was no doubt that DuBarry was happy for Marye's escape, but it left him seemingly alone in the rubble. He could only lie there in agony, waiting with hope for rescue. He was trapped in an enormous pile of timber. Only one thing could make his situation worse. DuBarry began to see sparks slowly drop from the firebox of the engine.

While DuBarry lay trapped, his friend and coworker, James Cassell, had dropped into Wolf Creek after his escape from the baggage car. The creek had carried him nearly three hundred yards away from the wreck until he reached a point where he could scurry up the eastern side of the washout and head back to the track. His head and hands were cut badly, his arm was broken, and he was losing blood. He stared into the darkness to get an idea of the circumstances, but he could only see the green glow of the lights on the back of the Calmar sleeper still sitting on the track. He decided that the best course of action was to head east toward the Thaxton station and get messages out to bring doctors to the scene as soon as possible.

On his way to the station, Cassell met John Read, the flagman from the freight train that had passed through ahead of passenger train Number Two earlier that night. Read had been waiting for the passenger train outside the station just a half mile away from Wolf Creek when he heard the train crash. His first thought was that the train had collided with something that had washed onto the tracks. He was on his way to investigate when he crossed paths with Cassell.

Cassell knew that there was yet another danger charging toward the wrecked passenger train. A freight train was only five miles behind when passenger train Number Two left Buford. If someone did not stop it in time, the calamity at Thaxton would be indescribable. The superintendent sent Read off on foot toward Buford to halt that train.

The rushing waters of the creek had carried Conductor Johnson about four hundred feet away from the wreck before he was able to swim to a tree and pull himself out. He scrambled up the high bank on the western side of the washout to the track where the Calmar sleeper was still sitting. As he scaled the hill, he noticed the embankment was slowly washing away. Johnson needed to act quickly, or Calmar would tumble into the abyss. He pulled himself up onto the platform of the rear sleeper and was met by one of Calmar's passengers, off-duty baggage handler J. H. Elam. Elam had dispatched the rear flagman, Samuel Boyd, to stop any trains headed east. Elam and Conductor Johnson stepped inside the sleeper, where Elam noticed that Johnson's skin was torn entirely from the left side of his face, leaving his left ear resting with a mass of skin under the cheek. In addition to the damage to his face, Johnson had sprained his ankle, and the steam had scalded him all over his body. The conductor collapsed into a seat to tend to his wounds.

Others who had freed themselves from the wreckage began to enter Calmar for refuge. Edith Hardester and Annie Fishpaugh had managed to climb out of the rubble with the help of L. H. Garnett and W. H. Craig, the Pullman porters on the Toboco sleeper. Annie actually had the presence of mind to grab her satchel as she escaped, carrying it with her onto Calmar.

J. H. Elam stepped outside to grab one of the lanterns attached to the sleeper. While Elam scrambled for the light, Johnson informed the passengers that the train had wrecked badly. He asked if anyone would go and help the engineer, Pat Donovan. There were no volunteers. Elam returned with the lantern and again asked the passengers for help rescuing others trapped in the wreckage. Despite the presence of at least eight men in the car, none offered to help.

There was not enough time to spend on a lengthy search for a Good Samaritan on Calmar. Conductor Johnson advised the passengers to get dressed and get out because the sleeper was in danger of falling as the embankment continued to deteriorate. Panic set in and many of the passengers began to scramble for the exits. Several of the ladies disembarked so quickly that they fled out into the rain still wearing their nightclothes. Elam jumped from Calmar with his lantern and slid down the bank about eight feet. As he did so, he immediately heard voices call out, "We are safe, there is a light!"[12]

One of those who welcomed the sight of that beacon in the darkness was Bishop Wilson. He was stuck in the mud so deeply that he was unable to get any kind of a footing to free himself. To make matters worse, his right hand had a match-sized wooden splinter run completely through his palm, rendering it temporarily useless. Nearby, a woman was crying out for her little girl. It was Catherine Thompson, Aunt Kate, and she was crying out for her beloved niece, Pattie. Wilson reached out with his left hand and helped Catherine pull free from the muck. They called for help, just about the time J. H. Elam arrived with his lantern. His first objective was to rescue Catherine, but she was in a hysterical state while looking for Pattie.

"Save the child!" she implored him and commanded him to let her go.[13] Elam held up his lantern and scanned the area for Pattie. There was no sign of her. He had to get Catherine to safety, so he told her that Pattie was safe and on the other side of the creek. He sold this tale well enough to convince her to come with him. He freed Bishop Wilson as well and escorted the two of them back up the embankment to Calmar.

While Elam's rescue efforts were just beginning, John Read made his way through the darkness on a mission to stop the oncoming freight train headed toward the wreck from Buford. When Cassell sent Read to stop that train, he was unaware that Elam had already sent the passenger train's flagman, Samuel Boyd, back for the same purpose. Flagman Read came upon the scene of the wreck and rather than attempt to cross the waters, he made his way precariously over the tops of the shattered cars. As he scrambled his way across the wreck, Read heard the screams of Catherine Thompson and others who needed help. He had to make a choice. He could stop and help those suffering in the pit below or continue on his way to protect them from the charging freight train. Read could do far more for the people trapped in the wreckage by protecting them from another onslaught, so he pressed on toward Buford.

After J. H. Elam assisted Catherine Thompson and Bishop Wilson to safety, he again made a plea with the passengers to help him with the rescue efforts. Once again his request fell on deaf ears. He descended the embankment alone once more to continue his attempts to free survivors. Those trapped in the wrecked Beverly sleeper were next to see Elam.

Inside Beverly, W. J. Barksdale and T. B. Bott worked together to make an escape route for themselves and the three ladies inside. Janie Caven, Inez Sparkman, and her mother, Roberta Powell, watched as the two men yanked a lamp from the car and used it as a battering ram to open a hole through the roof of the shattered car. Barksdale and Bott then hoisted three men up through the opening, only to have those men immediately flee the area once they had escaped. This lack of chivalry caused a bit of a panic for the ladies, one of whom turned to Barksdale and pleaded with him not to leave them stranded in the car. Barksdale's reply gave them all the reassurance they needed: "I will not leave, ladies, until I see you all safe; I am a Virginian."[14]

Outside the car, J. H. Elam and Burton Marye stood in water three feet deep. They had maneuvered themselves into position to assist with the rescue, and each of the ladies was lowered down to the men on the outside. Janie Caven was unhurt and actually had enough composure to change out of her nightclothes while the men were working on the rescue effort. Roberta escaped with some stiffness and soreness, but Inez seemed to be in the worst condition, with what appeared to be some sort of spinal injury. All three of the ladies were taken up the embankment where Calmar still waited. Inside Beverly, Barksdale had been true to his word and was the last one in the group to climb out of the mangled sleeper. Once he was free, he joined the other men to free the porter, W. H. Haywood. Haywood was not the only Pullman employee liberated by the impromptu band of rescue workers. His boss, John W. Scott, was also in need of their services.

John Scott was about thirty years old and worked as the Pullman Conductor on Beverly. His job was to manage the porter and the overall operations of the sleeper. When the train plunged into the washout, Scott had been launched through the roof of the car and pinned down by timbers from the wreckage. He was trapped, and despite his best efforts, he had found no means of escape. A "sickening, terrible feeling" had come over him as he listened to the wailing of those around him.[15] He could hear shrieks of terror from those who were hopelessly trapped mingled with the painful cries of the wounded. Scott could see small but ominous flares darting out from the engine's firebox. There was plenty of wooden fuel scattered everywhere to create an inferno that would easily burn him alive. He felt "helpless as a dead man."[16] Thoughts of his all but certain fiery death were abruptly canceled when he heard the sound of scraps of wood being torn away by J. H. Elam and the other men. He was rescued! They pulled Scott free, and despite being in a great deal of pain, he joined with the men to search for and free anyone else they could find. The rescue party continued to hunt through the ruins, as they called out and listened for the voices of anyone still alive.

Edmund DuBarry had been trapped in Beverly's wreckage for nearly an hour when he heard John Scott's voice calling for survivors. DuBarry summoned all the energy he could and cried out for help. His shouts reached Scott and Elam, and they quickly found him in the wreckage. His prospects were grim. In order to free him, they needed to do a lot more work than just pushing aside some debris. This job required axes and saws. Scott delivered the bad news to DuBarry. Edmund's hope was crushed, and fears of death began to overtake his mind. He asked Scott to let his wife and daughter know that his final thoughts were of them.

John Scott had been in DuBarry's position just a short while earlier. Trapped and waiting to be burned to death, he too had given up hope before his life was saved by heroic men. He knew more than anyone that DuBarry needed at least a drop of encouragement in his ocean of despair. Scott told DuBarry that they were working to get some other ladies and gentlemen free and up to safety. He assured DuBarry that once they got them up on the bank that they would return with the tools needed to get him free. DuBarry implored Scott, "As soon as you get them out, try and release me."[17]

Scott confidently replied, "All right sir. I'll be back."[18]

With that statement, John Scott returned to his rescue efforts, and DuBarry was alone again. He could only lie there and think about his mortality. He watched the threatening sparks that dripped from the firebox of the crippled engine onto the wooden timbers scattered everywhere. DuBarry had experienced many interesting things in his forty-six years of life. His father had been the personal physician and friend to Joseph Bonaparte, the King of Spain and brother of Napoleon. His uncle was William Duane, the former Secretary of the Treasury under President Andrew Jackson. DuBarry himself had been an eyewitness to the assassination attack on President Garfield on July 2, 1881. It was a terrible coincidence, but DuBarry now faced his own death exactly eight years later.

Wait. Wonder. Wait for help to arrive and wonder when the vast quantity of shattered wood from the railroad cars would catch fire. DuBarry was thankful that the sparks were held at bay as they licked at the wet lumber. The rain and the waters of Wolf Creek acted as friend and foe that night. First they had conspired together to cause the awful catastrophe, but they switched sides to become an ally to those who were still confined in the wreckage. As long as the timber stayed wet, passengers who remained trapped were safe from an all-consuming fire that would finish the job the water had started.

While the rescue work continued around the wreck, efforts were underway to bring additional help to Thaxton. The train's flagman, Samuel Boyd, ran nearly two miles west of the wreck and arrived at a place called "Bocock's Crossing." It was a road crossing named after the man who lived nearby, Benjamin Bocock. Boyd stopped and placed one of his signal torpedoes on the track to alert the engineer of the freight train that trailed behind passenger train Number Two. As he waited for the train to arrive so that he could signal it to stop, he saw a red light coming down the track from the direction of the wreck. It was John Read, the flagman from the freight train that had passed through earlier that night. Superintendent Cassell had sent Read for the same purpose, to stop the oncoming freight train.

Boyd was unaware of the extent of the damage to his own train because he had left almost immediately after the wreck. Read, on the other hand, had seen the carnage and heard the screams of the wounded as he maneuvered through the wreckage toward Bocock's Crossing. He brought Boyd up to speed on the terrible details. It was clear to both men that more help was needed at the wreck.

Boyd gave Read his signal torpedoes and flares so that he could wait by the track for the freight train. Boyd hurried to Benjamin Bocock's house and banged loudly on the door to wake him. Bocock, a farmer and a railway postal clerk, came to the door, and Boyd told him about the wreck near Thaxton. Bocock was quick to act. He immediately rounded up his farmhands and sent one man west toward Buford to get word out about the accident. He and the remaining farmhands joined with Boyd and rushed to Thaxton.

An injured and shaken James Cassell was at the Thaxton station working to get medical help to the crash site. In addition to all that had already happened, he was informed of another fill washed out east of the station. He had already sent flagman John Read to stop trains coming from the west toward the wreck, and he needed to stop anything coming from the east as well. Somehow he had to accomplish all of this without telegraph communications because the lines were down. The urgency to get instructions out to trains on the line was just one of his worries. He also needed doctors, and he needed them right away. The storms had not only attacked the train but also cut off the means of getting word out to those who could render aid to the dying and wounded. Cassell used the only resource he had available. He dictated a message and sent two riders on horseback off into the night. One traveled west to Buford and the other galloped east toward Liberty.

The injured passengers and crew who were able to get free from the wreck on the Thaxton side of the washout began to trickle in to the station. A man appeared at the door so battered and bruised that he looked like he had weathered seventy rounds of bare-knuckled fighting with the great John L. Sullivan. This man was no pugilist, however; it was postal clerk Lewis Summers. After his efforts to rescue Cassell and Ford, he had walked toward a red light in the direction of the Thaxton station. His head was cut so badly that he was nearly blinded by the blood running into his eyes. After his wounds were nursed, he collapsed and lay down to rest from exhaustion.

Another man who had survived the crash limped into the Thaxton station. Just before the wreck, Frank Tanner had left his friend John Kirkpatrick asleep in the first-class coach while he slipped up to second class for a smoke and a nap. Tanner's short attempt to get some shuteye had been violently interrupted by the train's plunge, and he had awakened to find himself on top of one of the wrecked cars. After jumping down into the water, he had helped some others get out of the wreckage but found the rescue work challenging in the "midnight darkness."[19] His ankle was sprained, but he decided to stagger his way to the Thaxton station to bring back a light. After assisting some of the wounded at the station, Cassell and Tanner hastily returned to the wreck together.

By 2:30 a.m., Edmund DuBarry had been trapped in the remnants of sleeper car Beverly for well over an hour. He heard many more voices around the wreck as word had spread of the disaster to the locals in Thaxton. The depot agent had awakened the section master, Tandy Jones, shortly after the wreck occurred, and Jones had immediately bolted to the scene. His watchman, John Johnson, also joined the rescue efforts. When the second culvert east of the Thaxton station had broken, the rush of water had burst through his home. The flood had ripped his doors off the hinges and blasted through his windows. The water had trapped Johnson and his family on a small upstairs stoop until nearly 2:00 a.m.

Passenger train Number Two's flagman, Samuel Boyd, had also returned to the wreck with Benjamin Bocock and several farmhands. Boyd spotted someone in the water as soon as he arrived. He rushed toward the figure hoping to lend a hand, but he was too late. It was Will Steed, one of the three friends from Cleveland, Tennessee. Steed's legs had been caught between some of the boards in the wreckage of the coaches. He had drowned in the waters of Wolf Creek.

DuBarry had already resigned himself to the same fate as Will Steed, but the sweet sound of axes and saws soon replaced sorrow with joy. John Scott had assured DuBarry he would be back, and he had been faithful in his pledge. He and J. H. Elam had returned with several men. They labored away with axe and saw to get the debris away from DuBarry. The men cleared a path and lifted DuBarry out of the crumpled sleeper car. Edmund DuBarry would be able to hug his wife and daughter once again.

DuBarry was worn out from his ordeal and suffered terribly from being scalded by the steam. He collapsed on some blankets spread out near the wreckage of Beverly. He reached for his glasses only to realize they were in his vest pocket, which resided in what was left of his sleeper berth. He asked John Scott to try to retrieve them for him. Scott's scavenger hunt not only yielded DuBarry's vest and glasses, but he was also able to find a lantern that had somehow survived the catastrophe in working order. Scott and the other men put the lantern to good use while they moved about the wreckage. They called out continuously for nearly fifteen minutes. No replies came from beneath the debris. The heartbreaking prospect that perhaps there was no one else still alive became very real, but the rescue party would not quit.

DuBarry recovered quickly from his hour of torment. He called John Scott and J. H. Elam together to find out what steps had been taken to get help to Thaxton. Elam informed him that flagman Boyd had set out toward Buford almost immediately after the wreck, but Boyd's job was to stop the freight train behind them. They already knew the telegraph was down at Thaxton, so someone needed to run all the way to Buford and get word to Roanoke to send help. Elam was the best man for the job. DuBarry implored Elam to "get the section force and a hand car, or if they were not available, to get a horse and wagon and proceed westward until he could communicate" if telegraph wires were also down at Buford.[20] Elam hurried off on foot to cover the six miles from the wreck to Buford.

DuBarry pulled out the watch from his vest and noticed that it was 2:50 a.m. Nearly an hour and a half had passed since the rain and floodwaters created a pit of death and destruction at Thaxton. Despite that, he had been thankful for all that water while he was trapped. The water had prevented the many shattered fragments of timber from igniting and burning him alive. Once Newman's Fill had completely washed away, the flood waters quickly receded into the farmland around Wolf Creek. The massive pile of timbers had been slowly drying out ever since. DuBarry glanced back at the spot he had spent that agonizing hour trapped in the sleeper, and his eyes beheld the sight he had feared the most. The sparks from the firebox were no longer going out. The wreck was on fire.

6

CONFLAGRATION

"The wrecked cars burned with a fierceness that could not have been equaled if every stick of timber had been soaked in oil."[1]

— HENRY N. MARTIN, PASSENGER ON TRAIN NUMBER TWO, JULY 4, 1889

"Can't you put that fire out?" DuBarry asked John Scott. "Those who are wounded will be burned to death!"[2] They tried in vain to scoop water and extinguish the fire. The train had collapsed on itself in the pit and the debris pile was thick. They could not get much water onto the source of the fire at the engine's boiler. The wrecked cars had assembled themselves into a makeshift woodpile over the engine's firebox, and a horrific bonfire was the inevitable result. The flames spread rapidly throughout the wreckage. The intense heat forced DuBarry and Scott to retreat. They stood together silently and watched the flames erupt. Rain began to pour, and DuBarry had only his shirt, underwear, and socks to protect him from the elements. Both men were thoroughly soaked, but the rain could not deter the blaze.

Suddenly, DuBarry heard something. Was someone groaning inside the remnants of the Beverly sleeper? Another member of the rescue party, Benjamin Bocock, thought he heard the cries of a man beneath the wreckage of the Toboco sleeper that rested on the side of the embankment. Were there still people trapped alive or was the hissing and popping of the steadily advancing fire playing with their minds? Several of the men moved closer to the blistering flames and frantically tried to locate those they believed had cried out from the rubble. The fire was relentless and worked with the determination of a starving predator happening upon wounded prey. Time had run out for any further rescue. The men backed away from the fire and climbed through the bushes on the embankment to the track where the Calmar sleeper waited.

They had cleared everyone out of Calmar earlier as Conductor Johnson had instructed. The worry at that point was that the embankment would give way further and swallow the last sleeper. Without the use of Calmar there had been no shelter from the rain for the wounded, and they had been placed on the track behind the sleeper exposed to the elements. An injured Bishop Wilson had already been there for about an hour and was lying in the dirt and stone track ballast with his head resting on the rail. The decision not to use Calmar for shelter had added to the misery of the survivors. As the fire scaled the side of Newman's Fill, it looked as though that decision might have saved lives.

DuBarry, John Scott, and some of the other men worked to uncouple Calmar from the shattered remains of Toboco. They hoped to free the rear sleeper and push it back from the wreck to provide a refuge for the wounded. The front platform of Calmar was jammed under the rear platform of Toboco, and they could not access the coupling to release the car. The men attempted to cut through the platform and attack the coupling with a sledgehammer, but their efforts were in vain. The fire continued its march through the wreckage. It was only a matter of time before it would reach Calmar.

Realizing they could not use the sleeper for shelter, they settled for the next best option. DuBarry instructed the men to strip the car of its contents. They grabbed mattresses, cushions, blankets, and anything else they could find to help relocate the injured passengers and crew further away from Calmar. The flames inched closer to engulf the last car. It seemed as if some dark force was determined to consume the entire train.

They retrieved what they could from the luxurious sleeper and moved the wounded further down the track until they reached a safe distance from the fire. Some passengers had earlier set out on foot to search for shelter in the sparsely populated area. Conductor Johnson and several other men and women from the Calmar sleeper had made their way two miles west to Benjamin Bocock's house for first aid and clothing. Others had journeyed only a mile or so hoping to stumble across a farmhouse or barn where they might find cover. The night was so very dark that they could easily walk right past a small house without even knowing it. Many passengers were in their sleeping attire, with bare feet, and little to protect them from the rain. Modesty was not much of a concern as they trudged across railroad ties and rock in the darkness, fruitlessly seeking sanctuary.

One such group of wanderers included Irene Jackson and her friends, Edith Hardester and Annie Fishpaugh. Irene had managed to grab her pocketbook, watch, and dress skirt before her escape from the broken rear portion of the Toboco sleeper. They walked a short distance and stopped to rest on some fence rails. Despite cuts that bled profusely, a few chivalrous gentlemen stood holding umbrellas to cover the ladies.

Those who chose not to walk away or were unable to do so rested about seventy-five yards west of the washout. Some were lying down on mattresses placed on both sides of the track, forced to dangle their feet in water that flowed steadily along the railroad ties. The survivors were at a safe enough distance not to worry about physical harm from the fire, but sounds and images began to permeate the scene that would leave mental scars forever. The burning wreckage illuminated the area like an early sunrise, and the fire produced a roaring sound that was unforgettable. Their eyes were fixated on the spectacle that had become more than just a washout and a terrible train wreck. The towering blaze that burst from the fill ensured that a little creek in a picturesque valley had become a crematory and final resting place for those who had not made it out with them. As they watched, some had already convinced themselves that all the survivors had been rescued and that the fire consumed only those who had already perished. It felt better to think that at least, but there was no way of knowing.

The sight of the flames brought a horrible realization to Catherine Thompson, little Pattie's "Aunt Kate." The fire would consume her beautiful nine-year-old niece. She had been pleading with the rescuers to bring Pattie to her from the moment J. H. Elam had first helped her to safety. As the inferno burned, she lay on the wet ground barely clothed and suffering from her injuries. She screamed with inconsolable grief. Tears flowed from her heavily bruised eyes, and her body convulsed violently as she wept. She lashed out at the men for rescuing her and not Pattie. "Oh, what will my baby think of me for leaving her?" she wailed. "Why did they not leave me and rescue my child?"[3] Her great sadness coiled itself around the hearts of those unfortunate enough to be nearby. Tears flowed from their eyes, and another cruel memory stamped itself forever into their minds.

The physical agony and emotional torture for the men and women of passenger train Number Two was far from over. It was three o'clock in the morning, an hour and a half had passed since the wreck, and they still had no idea when relief would arrive. They were far from any major source of help, and telegraph communication was out. Their hope was that men like J. H. Elam traveling on foot or Cassell's horseback messengers could get word of the accident to the nearby cities of Buford, Liberty, and ultimately Roanoke. The weather meant that the night was fraught with many dangers, and men on foot or horses could easily be delayed, or worse. Even if one of the men reached Buford or Liberty, how could he be sure that the telegraph was not out further down the line? Help would eventually arrive, but it would not be soon.

J. H. Elam was about a mile and a half into his trek to Buford for help when he noticed a bright flash of light radiate from behind him. He turned and saw the sky illuminated in the direction of the wreck. He was standing on the downslope of a hill and could not see the source of the eerie light, only the glowing sky. Elam did not need to see the flames to realize his charge to get help had just been escalated to new heights.

The wreck was on fire, and he would run all twenty miles to Roanoke if he had to.

As Elam soldiered his way through the darkness toward Buford, one of the horseback riders dispatched by Cassell finally arrived at the Buford station with news of the wreck. After two and a half hours of suffering, someone outside of Thaxton finally knew the fate of passenger train Number Two. It was just shy of 4:00 a.m., and the telegraph operator quickly transmitted a message to Norfolk & Western's General Manager, Joseph Sands, in Roanoke. Sands wasted no time. He ordered his men to assemble a relief train and to summon the town's physicians from their peaceful sleep. Drs. Joseph Gale, Richard Fry, Arthur Koiner, Herman Jones, and R. Gordon Simmons dressed hurriedly and headed for the station.

While N&W mustered help in Roanoke, J. H. Elam had nearly completed his trip on foot to Buford. Just off in the distance, he could make out the joyous shape of a locomotive coming toward him from the station. He desperately flagged the engine down, and it slowed to a halt. Cassell's horseback messenger had already delivered the terrible message about the wreck, but Elam brought word of the horrifying fire at Thaxton.

Elam hopped aboard the engine, and they immediately backtracked to the Buford station. At Buford, Elam's news was sent over the telegraph to the N&W general manager. At 4:30 a.m., the fire alarm bell echoed throughout the slumbering city of Roanoke, and the Vigilant Steam Fire Company responded. The firemen and their steam-powered fire engine were added to the relief train, and they sped off from Roanoke toward Buford around 5:00 a.m. Three and a half hours had already passed since the crash, and the fire had been burning for two hours.

While the gears slowly began to turn in the rescue effort machinery, the passengers and crew remained stranded at Thaxton with no idea when help would arrive. The worst of the rain had finally subsided around four o'clock, and some of the victims who could be moved were taken to nearby farmhouses or to the station at Thaxton. The rest lay scattered about alongside the tracks, resting on a sea of mattresses.

Several ministering angels tended to the wounds and needs of the injured. Pauline Payne had been with her brother Reuben on the last car, Calmar. Similarly, Rosa Lee Hurt and her brother John Irby had been in the Toboco sleeper when it ripped apart during the crash. Pauline and Rosa Lee were uninjured, and they had not taken their good fortune for granted. Instead of heading for cover, they stood in the rain with only blankets covering their nightclothes and provided help to the others. Pauline had seen to it that Bishop Alpheus Wilson had a mattress to lie on, and she had tied up his damaged hand.

Fred Temple had dragged himself to a sleeper after he escaped the wreckage of the first-class coach. Rescuers had found him there and placed him alongside the tracks next to Bishop Wilson. Temple was badly injured, and he watched as the men and women who were willing to help scurried about doing everything in their power to lend a hand. He saw a stark contrast between those who were "doing all in their power to relieve the suffering, while others stood by totally unconcerned about anything but their own comfort, cold and heartless as it were possible for human beings to be."[4]

A mud-covered Janie Caven was most certainly not one of the selfish ones. She drifted from person to person, caring for the injured as best she could. The young college student had been trapped in Beverly after the wreck, and from the moment she was rescued from the sleeper, she had acted heroically. Janie had labored side-by-side with the men as they pulled trapped passengers and crew free from the wreckage. She had escorted at least one injured passenger to a nearby farmhouse, and then returned to the accident to continue the work. Once the rescue effort was finished, she began to rip strips from the fabric of her dress to use as bandages for the injured. A few of the other ladies joined in to offer much needed comfort to the otherwise helpless men and women who suffered through the night with no medical care.

The kindness offered to strangers by the ladies was encouraging, but Edmund DuBarry began to get anxious. They had been waiting a long time for help, and there were still no signs of it. People needed attention, and they had been forced to endure the night perched above a smoldering mass of death. He and James Cassell discussed their options.

Was it possible that some other calamity had taken place on the road to Buford and cut them off from any chance of getting help? They knew the weather had already caused two washouts on either side of the Thaxton station, and it would certainly be possible that the storms had inflicted more damage elsewhere. The superintendents came to the decision that they would go to Buford themselves. Both Cassell and DuBarry had gone through traumatic events that night, and they were not far away from Thaxton before they realized one of them should probably stay behind with the wounded. Cassell returned to the wreck, and a battered, bruised, and burned Edmund DuBarry set off on his own six-mile hike west to Buford.

Even though at Thaxton the passengers remained unsure whether anyone knew of their status, news of the disaster was slowly spreading. Around 5:00 a.m., a messenger on horseback arrived at the depot in the town of Liberty, five miles east of Thaxton. Freight train conductor Butler, the man in charge of the last train to pass over Wolf Creek safely that night, greeted the messenger. Butler had been at Liberty since midnight, as he waited on conditions to improve and for his flagman, John Read, to return from Thaxton. Butler had left Read behind to warn passenger train Number Two of treacherous conditions on the road. He and cattle dealer T. P. Ayers had been waiting at Liberty for nearly five hours, but there had been no sign of Read or train Number Two. For reasons unknown, Cassell's messenger had taken two and a half hours to travel just five miles.

The message from Cassell was chilling. Number Two was badly wrecked, and the survivors needed all the physicians they could get. Butler pulled his engine from the siding and hooked up to a caboose while Liberty's doctors were summoned to the station. Physicians Thomas Bowyer, David Wade, and Walter Izard hurriedly climbed aboard the caboose and the engine sped off to Thaxton.

Despite all he had been through, DuBarry had nearly completed his trek to Buford. The sound of escaping steam had tormented him earlier that night, but the sound of an approaching engine suddenly became music to his ears. A relief train was coming toward him, and on board were Buford Doctors Samuel Price, William Rice, and William Spinner.

DuBarry climbed aboard, and the train throttled up for the rush to Thaxton. Upon arrival, the Buford doctors went straight to work alongside those from Liberty. The wreck survivors had spent four hours' of agonizing isolation in the angry, dark night, but help had finally arrived. Cassell immediately ordered the engine back to Buford to retrieve two cabin cars on a siding there to use for temporary shelter at the crash site.

The train from Roanoke that carried N&W personnel, doctors, and firemen pulled into Buford at nearly six in the morning. General Manager Sands was informed that the need for fire suppression had long passed. The fire had ignited in full force nearly three hours earlier, and it had done its handiwork well. Everything, including the Calmar sleeper that had remained on the track, had been consumed. Sands decided to leave the firefighting equipment at Buford. Manpower was still needed, so the twenty firemen stayed on the relief train, and they sped off to Thaxton. The doctors and firemen from Roanoke arrived at Wolf Creek around 6:25 a.m., which was exactly five hours after train Number Two had met its demise.

The physicians did what they could at the site, and several hours passed while they tended to the many wounded. Postal Clerk Lewis Summers and five others were already at the Thaxton station receiving treatment. The men at the station were taken to the newly constructed hospital in Liberty. Nearly all the remaining passengers and crew, the injured and uninjured, welcomed the chance to board a relief train bound for Roanoke. One single heart-rending exception was Catherine Thompson. Her overwhelming sorrow for her precious niece could be felt by all present. She knew Pattie had perished and was somewhere among the ruins. In her heart she felt that she had abandoned Pattie when she was rescued from the wreckage, and she could not bring herself to leave her again. It took an effort of "great difficulty," but they were finally able to coax Aunt Kate onto the relief train.[5]

The subject of where to take this many wounded had never been discussed before. Roanoke was a new city and did not yet have a hospital facility. Dr. Gale suggested they use Hotel Roanoke as a temporary hospital. There were nearly fifty rooms there, and along with the

City, Felix, and Palace hotels, they created impromptu medical facilities all around the city. The relief train loaded with doctors and survivors set out west toward Roanoke near ten o'clock on the morning of July 2. As it pulled away, smoke continued to drift into the sky from the depths of the washout. With the survivors all safely carried away, the fire equipment left at Buford was called for once again. The scorching hot ironwork snarled together down in that hole needed to be cooled before the men could begin the next gruesome task.

7

AFTERMATH

"Honor thy father and thy mother; that thy days may be long upon the land which the Lord thy God giveth thee."[1]

— GEORGE RUTLEDGE STUART, QUOTED FROM THE
BIBLE IN A SERMON AT ST. LOUIS, MISSOURI, MARCH
8, 1895

By the time night had fully yielded to day, knowledge of the wreck was widespread. Thaxton's population ballooned with the arrival of men, women, and children determined to see the devastation with their own eyes. Townspeople from nearby Liberty first became aware of the disaster around five in the morning, and they rushed to the scene. The day's work was abandoned, and they went on foot, horses, and buggies to take in the spectacle.

While some gathered at Thaxton out of simple curiosity, others were there to offer help. Unfortunately, as is often the case when humans gather, some were there with more sinister purposes in mind. The chaos at the scene offered opportunities for the criminally inclined to profit from the misery of others. Early that morning one of the passengers, Pauline Payne, had her satchel stolen along with the money and jewelry inside while she had been busy tending to the wounds of others.

Criminals, do-gooders, and onlookers alike all gawked at the same sight. An unrecognizable pile of iron and rubble lying at the bottom of a roughly eighty-foot wide, twenty-three-foot deep gash carved through the railroad embankment. It looked as if someone had carelessly swiped an eraser across a pencil sketch of Newman's Fill. The fire had burned so fiercely that the unconsumed fragments looked nothing like a train. The snarled mixture of iron, wheels, rails, and debris looked as if some great bird had weaved together a nest using the leftover pieces of train Number Two. A careful observer might have been able to spot the engine's colossal driving wheels or steam dome in the wreckage, but even that was a challenge.

The fire had destroyed everything, which included the magnificent rear sleeper, Calmar. Only the wheels of Calmar remained, perched at the edge of a steep drop on the western side of the washout. Across the gap, the demolished remains of mighty engine Number 30 were lying on the washout's eastern slope. Smoke continued to rise from the smoldering heap, and firemen soaked everything they could with hoses to cool the scorched metal. Men wearing fire helmets and jackets carried the hoses down the embankment with the help of local citizens who volunteered to help. A number of onlookers stood on the hill above the engine and watched the men at work down in the pit.

Spectators also watched from positions on the ground scattered about the newly formed muddy flood plain. Some of them sat on very large rocks that were likely carried down the creek and through the washout by the rushing waters the night before. Residents throughout the area reported rocks weighing several tons and trees forty feet long were carried down the mountains by the floods.

Streaming out from beneath the wreckage was Wolf Creek, once again looking more like the small country stream it had been. It was hard to believe the harmless little creek could ever have been powerful enough to tear a massive hole in the earth. The storm that pounded Bedford County the night before was devastating in its power. Many of the locals attributed the sudden surge of water through Wolf Creek and other creeks in the area to water spouts or cloudbursts. There was telltale evidence of several cloudbursts on the ridge above Thaxton as well as surrounding areas. An outbreak of flash flooding occurred throughout that section of the country because so much rain had fallen in such a short amount of time. The second washout that occurred near watchman John Johnson's house on the east side of the Thaxton station had created a nearly two hundred-foot wide, sixty-foot deep hole in the embankment there. Seventy-year-old James Chilton lived nearby that washout, and his stable and shed were carried away nearly three miles by the flood. Chilton had never seen anything like it.

Thaxton was not the only town victimized by the storm. George Nichols lived in the neighboring town of Liberty. He walked outside the morning after to survey the damage around his farm. The water had carried off four hundred panels of George's fencing. He checked a post that marked the previous high water mark set during the "big flood" of 1870, and there was no contest.[2] The downpours the night before had surpassed the 1870 record by four to five feet.

Another man named John Toms lived on Sharp's Mountain above Thaxton. John knew of at least four "water spouts" no more than a half mile from his house. His mother lived next door, and between the two of them, they figured they had about a hundred dollars in damages each. Their losses included one plow that "has never been since heard of."[3] In fact, there were over eleven washouts found on Sharp's Mountain and nearby Suck Mountain. If there were any question about the brute strength of the floods, the 50,000-pound granite boulder carried down the side of Suck Mountain for a quarter of a mile removed all doubt.

Rivers throughout Virginia had far exceeded high-water mark records. Vast quantities of crops were destroyed, and several grist mills were completely washed away. Bedford County estimated between $25,000 to $30,000 in damages just to roads and bridges ravaged by the storm.[4] Floodwaters had completely carried away an iron truss bridge at Wilkes' Mill on Big Otter Creek. The storm also knocked a three hundred-foot-long bridge over the Little Otter River nine miles east of Thaxton nearly a foot out of line. The total damage to crops, livestock, and property in the county soared beyond $100,000.[5]

The deluge of water had wreaked havoc all along the east coast. Washouts and landslides crippled the Virginia Midland Railroad near Lynchburg and the Richmond and Alleghany Railroad near Wilton, Virginia. Trains scheduled to run from Baltimore to Washington, DC, on the Baltimore & Potomac railroad sat idle due to washouts in that area. Even as far away as suffering Johnstown, new flooding from the heavy rains had turned up bodies that had been buried in the mud for a month.

With transportation at a virtual standstill all over the state of Virginia, it was not a surprise to see a crowd gathered at the Lynchburg train depot thirty miles to the east of Thaxton that morning. These were not stranded passengers waiting on things to get moving again, however. Passenger train Number Two would typically have arrived at Lynchburg around one thirty that morning, but instead it had been swallowed by Wolf Creek at about the same time. A sign hung up at the depot simply stated, "No Trains for the West To-Day."[6] An explanation was not necessary because just about everyone in town had heard about the terrible wreck at Thaxton. Concerned citizens gathered around the depot and the newspaper offices, hoping desperately for reports from the accident scene. Fears over potential lost loved ones gripped friends and relatives who had not yet heard any details. Some knew for certain that they had a loved one on the train, but there were others who could not be sure whether someone they knew had taken an unscheduled trip. Until more news arrived, they could only wait and worry.

The situation was much the same to the west of Thaxton at Roanoke. A multitude gathered around the depot and congregated in the streets, while rumors based on fact and speculation spread throughout the city. The people of Roanoke were slightly better informed than the Lynchburg residents because the washouts forced nearly all rescue and recovery efforts to originate from Roanoke. The clanging alarm bell that summoned Roanoke's firemen in the wee hours that morning was the first indication to everyone that trouble had developed somewhere down the rails. Roanokers began to see for themselves just how terrible the wreck had been as the injured arrived on the relief train. When Bishop Wilson was escorted from the train, he was so covered with blood and mud that his own family would not have recognized him. It was clear from the news around town that death had visited the train, but they had no idea how many had died. Sadness took hold of the city. Men normally known for their toughness were openly weeping in the streets. A number of women wandered the streets, hoping to find someone who could tell them their loved ones were safe. It was certainly not a situation in which "no news is good news." Just like their counterparts in Lynchburg, they needed more information to answer the question of the hour. How many had been lost, and who were they?

The first confirmed loss of life at Thaxton was Will Steed. Brakeman Samuel Boyd had pulled Steed's body away from the creek in the darkness before the fire made its assault. It was not clear whether Steed was in the first-class coach or in the smoking car when the train went down. One theory was that he and his two friends were in the smoking car, and Steed may have attempted to jump through the window at the moment of the wreck. There was no speculation necessary to determine what had caused Steed to take his final breath. He had suffered three broken ribs and a serious wound under his right eye, but Wolf Creek had drowned Will Steed.

Word of the young man's fate reached Cleveland, Tennessee, before the relief train with the survivors had departed Thaxton. A telegram arrived in Cleveland just after 9:00 a.m. with a shocking announcement: "Party bearing your letter of introduction to E. Townsend, New York City, was killed in a wreck—name W. C. Steed."[7] Steed was carrying the letter of introduction given to him by someone at the Cleveland National Bank before he left for his trip. Railroad officials found the letter on Steed's body and sent a telegram to the Cleveland National Bank to notify them of the situation and to determine what to do with his body. News of Steed's demise spread quickly. The citizens of Cleveland began to mourn the loss of one of their promising young men. The whereabouts of his two traveling companions, John Hardwick and Will Marshall, was a mystery. Their fathers, Christopher Hardwick and Samuel Marshall, decided they would leave on the next train for Roanoke and find their sons.

The task of getting to the wreck turned out to be more of a challenge than it should have been for those who lived closer to Thaxton. The railroad officials for Norfolk & Western had already switched into corporate damage-control mode. Management instructed employees in Lynchburg to pass on no information concerning the wreck and to prevent the press from traveling to Thaxton. When asked for details, the employees would only reply, "they knew nothing," which prompted the newspapermen to describe them as "dumb as oysters."[8]

Crowds of concerned friends and family lingered around the train station at Lynchburg, hoping to jump on any salvage or relief train headed west. The railroad employees worked hard to limit access to the scene to only those who might be friends or family of someone on the wrecked train. Some newspaper reporters attempted to jump on the platform of a passenger coach headed to the wreck and were blocked immediately by N&W employees. They told reporters it was against company orders for them to be on the train. One railroad employee made the company's position crystal clear to one of the reporters: "If you show me a newspaper man on the train I will have him put off."[9]

The decision to limit access to the press was at the very least a tremendous public relations miscalculation. The desire for details of the accident and casualty names was intense throughout the region. Newspapers used descriptions like "tyranny," "outrageous," "an act of official discourtesy, as rude and senseless as it was useless and uncalled for," "a wrong that was inexcusable," and "a piece of unmitigated stupidity" to characterize the decision to block reporters. Even Virginia Railroad Commissioner James C. Hill would later describe the decision as "idiotic."[10] If it seemed as if they mismanaged the handling of the wreck, it was likely because Norfolk and Western had never dealt with anything quite like it. This was the most serious accident that had ever occurred in the state of Virginia, much less on the N&W railroad.

Jacob Swaim, an agent for the Travelers' Insurance Company, was one of the few who successfully secured a ride from Lynchburg to Thaxton. Upon arrival he spoke with Colonel Frank Huger, the Superintendent of Transportation for Norfolk & Western. Huger had served as a colonel for the Confederacy during the Civil War and was the son of Confederate General Benjamin Huger. Frank Huger had been working at the wreck since he arrived early that morning with General Manager Sands from Roanoke. He informed Swaim that their best guess was twenty people had been killed. Swaim was particularly interested in the status of three men covered by his insurance agency. The men were fireman James Bruce, conductor Roland Johnson, and a brakeman named W. C. Glass. The news for one of Swaim's clients, twenty-two-year-old James Edgar Bruce, was not good. Bruce was gone, killed while he worked at his post shoveling coal into engine Number 30.

Swaim's other clients had fared better. Conductor Johnson was alive, but he was in bad shape. Johnson was taken to Roanoke to get treatment for his many cuts and bruises, a sprained ankle, and the scald burns he suffered all over his body. Brakeman W. C. Glass was riding on the second-class coach when the wreck occurred and had suffered scald injuries to his face and a broken left arm. He was one of six men who were taken to Liberty's newly opened hospital, Granville Sanitarium.

Five of those six men taken to the hospital in Liberty were crew members. Doctors treated Postal Clerk Lewis Summers, Baggage Master William Ford, Express Messengers William Graw and Robert Ashmore, and Brakeman Glass for a range of injuries including scalds, concussion of spine, head injuries, and numerous cuts and bruises. The sixth man at Liberty was passenger Robert Davis. Davis was the black Methodist minister relegated to a seat in the second-class smoking car because of his race. As it turned out, racism may have saved his life that night. More lives had been lost in the first-class coach than were taken from the rest of the cars combined. There were no deaths reported in second class. In order for Robert to receive treatment for his injuries, Granville Sanitarium had to make an exception to its own rules on race. For that moment at least, racial segregation was cast aside during a time of crisis. Robert was burned on his face and hands, and he suffered a slight concussion. All six men were especially lucky to be alive. The mail car, baggage car, and second-class coach were closer to the engine and its deadly fire than any of the other cars.

Recovery work continued at Thaxton. As workers sifted through the charred remnants of engine Number 30, they discovered the dreadful fate of the men on board. All that was left of Engineer Patrick Donovan was his torso, but his family would at least have some evidence of his end. Only fragments of bone were found for the other two men, James Bruce and Alvin James. Even though he was not working that night, Alvin volunteered to ride on the engine from Buford to Thaxton to ensure careful operation of the train. His gesture to help had cost him his life and spared Superintendent James Cassell from certain death.

Statistically, the most dangerous job on the railroad was that of the postal clerk. Sadly, statistics held true once again. Lewis Summers had escaped with injuries that he would soon recover from, but James Rose would never make it to the altar with his sweetheart, Lillian. Rose's remains were positively identified, but that would prove to be a rare circumstance once the recovery efforts were concluded.

Workers scoured the wreckage for remains and found that the fire had destroyed almost all trace of human life. With nothing but ashes and bits of bone to go on, officials could not calculate with any accuracy the number of people who had been consumed in the flames. Efforts to determine how many had perished were further complicated because Conductor Johnson's ticket records were stored in the second-class coach and had vanished in the fire. How could railroad officials ever be sure how many had died? They had no official record of exactly who was on the train, and in most cases, little was left to identify those who were lost. Officials could rely only on the inquiries of friends and relatives over the next several days to reveal some of the names.

Christopher Hardwick and Samuel Marshall were not looking for remains. They departed Cleveland, Tennessee, around noon and were bound for Roanoke, hoping to find their sons alive and well there. If necessary, they would continue from Roanoke to Thaxton to search for their boys. All of Cleveland held out hope for John Hardwick and Will Marshall. They already knew Will Steed's body had been recovered, but conceivably his friends could have survived. Norfolk & Western's stranglehold on information released from the scene of the wreck not only angered the press, but left Hardwick and Marshall's families with terrible uncertainty. The optimist could only hope that they had been taken to Roanoke with the other survivors.

Aftermath

The concern of citizens in Cleveland was just as apparent on the streets there as the scenes that played out in Lynchburg and Roanoke. Three well-known Cleveland men from prominent families were on passenger train Number Two. One was already confirmed dead. The telegram announcing Steed's death had sparked the search for his traveling companions, and that same telegram had triggered a search for another man, John Stevenson. Stevenson was from Richmond, and he and his boss, John Bowers, had been traveling on business during the week leading up to the wreck. They had completed their business, and the men split up at Bristol, on the Virginia-Tennessee border. Stevenson headed for his home in Richmond, and Bowers headed to Cleveland on further business.

Bowers found himself in the midst of a grieving town once news of Steed's death arrived. John Stevenson had worked for him since he was a boy, and Bowers was fond of him. Fearing that Stevenson may have been on passenger train Number Two, Bowers made his way to the telegraph operator and sent a message to his bookkeeper in Richmond. It was brief and to the point. "Where is Stevenson?"[11]

The nature of travel and communication at the time often meant that friends and family could not be certain when a traveling loved one would be returning home. The telegraph was the primary means of long-distance communication, and a traveler wouldn't necessarily send a message to let someone know he or she was on his or her way home. John Stevenson's wife, Ada, was unsure where her husband was or when he would be returning. She had not heard from him, and her fear was that he had fallen ill somewhere in his travels. She had no idea that he might be on the train and had been asking around town if anyone knew of his whereabouts. No one in Richmond had heard from Stevenson, but the telegram from Bowers raised fears that he may have been on the wrecked train. Information was slow in coming. Family and friends could only wait and pray.

The telegraph wires were highways of sadness as information leaked slowly out from Thaxton. A telegram sent to Lynchburg around 11:00 a.m. revealed the terrible news that John Kirkpatrick was a victim of the wreck. The city grieved over the loss of the popular young man, who was just in his mid-twenties. Kirkpatrick's fate was especially difficult for some to accept because he had never planned to be on passenger train Number Two. Because he unknowingly endorsed bad checks written by a friend, Kirkpatrick had traveled to Roanoke and lost his life.

The announcement of Kirkpatrick's death spread a cloud of gloom over Lynchburg. Townspeople would not greet one another with "How are you?" or "Good afternoon!," but instead the question of the day was, "Have you heard anything later from the wreck?"[12] Most of the time the question was from someone concerned about friends or relatives, but others had more practical matters in mind. Merchants around town were expecting checks from other regions to be delivered on the passenger train, and loss of revenue was foremost in their minds. Businesses would never receive that revenue because the entire train and its cargo were destroyed. Sixty packages of letters, fifteen sacks of papers, three registered mail pouches, the clothing and personal effects of the passengers, and an express safe containing thousands of dollars in treasury notes, jewelry, and watches had all been devoured by the fire.

Even as the dead remained undiscovered in the wreckage and the recovery work continued at Thaxton that afternoon, some of the uninjured passengers continued on to their destinations. John Rowntree, the hardware buyer from Knoxville, had survived the wreck in the Beverly sleeper. By 2:00 p.m., he had already left Roanoke and resumed his trip to New York. The business of moving people and freight would not pause for the dead at Thaxton. Norfolk & Western made arrangements with the Richmond & Alleghany Railroad and the Shenandoah Valley Railroad to allow trains to bypass the route running through Thaxton. The freight train that had been the last to successfully clear Wolf Creek finally resumed its trip around 4:00 p.m. after the resolution of issues with the Little Otter Bridge east of Liberty.

Onlookers dwindled as the spectacle at Thaxton became more of a worksite. A wrecking train equipped with a crane apparatus had replaced the fire-fighting equipment used earlier in the day. As nightfall approached, construction had already begun on a trestle to restore the railroad's path over Wolf Creek. As the men worked that evening, they pulled from the rubble a portion of a small body and carried it to the undertaker hired by Norfolk & Western. Pattie Carrington had entertained the passengers on the Toboco sleeper just the day before, but she would be entertaining angels forevermore. Pattie's death was communicated via telegram to her father in the midst of his honeymoon trip to Europe. He had lost her mother during Pattie's birth, and now the wreck had snatched his nine-year-old daughter away.

A telegram was the only way to get the devastating news to Pattie's father overseas, but Cleveland, Tennessee, was just a day trip away. Around midnight, a train steamed into Roanoke carrying Christopher Hardwick and Samuel Marshall. No one had seen or heard from their sons John and Will since the wreck. The two fathers made their way from the train station to undertaker George Sisler's office. What were they truly hoping for as they walked through Sisler's door? It was possible that their sons were still alive and well, but that seemed to be a very dim ray of hope. They were already aware that their sons' friend, Will Steed, was dead, and they had received word that there was no sign of their boys at the wreck. It was an emotional struggle that would rip any parent apart. If their sons were not in Sisler's office, then the likely alternative was even worse.

The undertaker could not offer them any closure. The body of Will Steed was there as expected, but John Hardwick and Will Marshall were not. A Steed family friend had also made the trip to Roanoke, and he boarded a train to escort Steed's body back to Cleveland. The two distraught fathers remained and, as any loving father would, these men were willing to turn all of Virginia inside out to find their boys. A father would stop at nothing to reach into the wreckage and bring home whatever might be left of his son. They left Roanoke for Thaxton as soon as they could on the morning of July 3 and then combed the debris, desperately searching for any evidence of their boys. Marshall offered $1,000 to anyone who could help him find Will's body. Their search for some trace of their sons to hold on to would continue all day.

While Marshall and Hardwick searched the wreckage that morning, news of the disaster was in newspapers from coast to coast. The stories had varying degrees of accuracy with just about all of them being incorrect in some facet. Many articles reported that the train was going at a high rate of speed and leaped into an open chasm in front of it. These accounts were often based on speed estimates given by passengers who were actually asleep when the train wrecked. The train had taken twenty-five minutes to cover six and a half miles from Buford to Wolf Creek at Thaxton. The slow pace meant the trainmen had proceeded with caution at least to a certain extent. The manner in which Calmar was left on the track, and Boyd and Elam's descriptions of only a slight jolt to Calmar also seemed to dispute the notion of a high-speed leap. The position of the cars in the wreckage provided additional clues that indicated the fill had collapsed while the train was on it.

The number of dead and injured was also inaccurate in most articles. In some cases, the names listed included people who were not even on the train. A higher body count printed often meant more papers sold. Some papers stretched the number killed to forty or fifty and would cite unnamed "informants" as the source for their information. The N&W decision to block the press from Thaxton did not make things any easier, and, as a result, rumors and stories relayed through anonymous sources were the norm. Some people even accused the railroad of combining the remains of multiple victims into one box to give the impression that fewer lives had been lost.

The reality was simply that no one knew how many people had been incinerated in the fire. At least ten people were already known to be dead by the morning of July 3. James, Donovan, and Bruce had all been killed on the engine. Rose had perished in the mail car, but the first-class coach was where most of the carnage had taken place. It was believed that the deathblow was delivered when the coach was crushed by the Beverly sleeper. Nathan Cohen was confirmed dead. He would never make it to see his parents again in Germany. No trace of his body was found, but his coworkers in Roanoke recognized Cohen's felt hat that had been retrieved from the wreckage. The two Norfolk & Western employees who were on a "pleasure trip," James Lifsey the dispatcher and Dennis Mallon the janitor, were both crushed and burned in the coach as well.[13] Lifsey's brother arrived in town to bring his brother's remains home, only to learn the shocking news that there was nothing left for him to take. There was no real trace of Mallon either, but Lifsey's family would at least receive some proof of his demise. Workers sifting through the debris would later find his blackened double-case gold watch and send it to his family in Greensville County, Virginia. John Kirkpatrick and Will Steed had also been confirmed victims in the coaches. Young Pattie Carrington was the only one known to have died in the sleepers.

The total number killed in the first-class coach was certain to grow. It had become a foregone conclusion that John Hardwick and Will Marshall had likely met the same fate as their friend Will Steed. While a telegram had announced the wreck and Steed's death in Cleveland, most people around the country knew nothing of the situation until the newspaper articles were printed the next day. As more people heard about the wreck, more potential passengers were identified. George Peyton was a telegraph operator for the Virginia Midland Railway living in Alexandria, Virginia. After learning of the wreck, he set out for Thaxton on Wednesday morning, July 3. He knew his brother Charles had been on the train with his wife and baby girl. There had been no communication from Charles, and they were riding on the doomed first-class coach where so much death had occurred.

The newspapers announced the wreck in Knoxville, Tennessee, as well. An employee of the Union News Company in Knoxville was particularly troubled by the news. He composed a disturbing telegram to Augustus Becker in Richmond, Virginia. The telegram stated, "Mr. Wheeler left here on the train that was wrecked on the Norfolk and Western road."[14] Becker was the Superintendent of the Southern Division of the Union News Company, and Mr. Wheeler was Harry B. Wheeler, one of Becker's employees.

Wheeler was not just any employee to Augustus Becker. He was one of his favorites. Becker had asked Wheeler to come work for him as his assistant in Richmond about ten months earlier. Wheeler had not hesitated to leave his home in Baltimore with his wife and five-year-old son and had moved to Richmond. They had tremendous admiration for each other, and Harry had been working for Becker for many years. As a matter of fact, Harry liked his boss so much that he had given Becker's name to his little boy, Arthur Becker Wheeler.

Becker composed a telegram to Norfolk & Western General Manager Joseph Sands and to Mr. Wright, one of the railroad's agents at Lynchburg, to find out whether they had seen Wheeler. Wright traveled to Thaxton to inquire about Wheeler but was unable to find him. Becker received a reply from Sands via telegram that was even more distressing. "We have no knowledge of Mr. H. B. Wheeler whatsoever. If he was on the train that was wrecked he has not been accounted for, and I presume has been lost. His body has not been recovered, but may have been burned in the wreck."[15] Like many of the friends and family of passengers on the train, Becker departed Richmond for Thaxton to search for his friend himself.

Some family members received happier news from the wreck. William Caven received a telegram in Dallas from his teenage daughter, Janie. She had acted heroically after the wreck by lending aid to the wounded, and she then reached out to her father for help. The telegram simply stated, "I was in the Thaxton wreck. Am not injured. Come to Roanoke. JANIE CAVEN."[16] Janie's friend, Inez Sparkman, and Inez's mother, Roberta Powell, had also managed to get a telegram back to their relatives in Texas. They were not seriously injured, but they had lost all of their clothing. The ladies of Roanoke supplied them with new clothes and ample hospitality.

Even as friends and family streamed toward Thaxton to search for their loved ones, some victims and survivors of the wreck were going home. The remains of the train's engineer, Patrick Donovan, arrived in Lynchburg on Wednesday morning. The life of an engineer in the nineteenth century was dangerous, exciting, and even a bit glamorous. They were the captains of giant fire-breathing dragons that could turn on them at any moment. Engineers were familiar faces to the people in the towns on their runs. The style they used blowing the train's whistle could be unique to each man. Donovan had made his last run, and his body was met at the station in Lynchburg by a large group of the town's Irish population. Services and burial were conducted that very morning at the Church of the Holy Cross where Donovan was laid to rest.

Will Steed's body finally made it home on an evening train into Cleveland on July 3. The city came out in droves to show their love for Steed. Two thousand people were at the depot to meet the train carrying his body as it pulled into town. Funeral services were scheduled for the next day. There would be no happy Fourth of July celebration in Cleveland.

Some of the fortunate survivors continued on to their destinations the same day that trains carried the remains of Steed and Donovan home. Annie Fishpaugh and Edith Hardester left Roanoke for Baltimore on Wednesday morning. Henry Martin, one of the passengers on the Calmar sleeper, arrived in Washington, DC, on the third as well. He had even managed to save his suitcase before leaving the train. Bishop Wilson departed Roanoke on the third, but all he had left were the cuff buttons that had been attached to his nightclothes. He headed for his sister's house in Charleston, West Virginia, hoping to make it back to Baltimore on the fourth. Postal Clerk Lewis Summers had checked out of the hospital in Liberty by Wednesday afternoon and had made his way to a hotel in Lynchburg known as the Lynch House. Summers was described by others in Lynchburg as "considerably cut and bruised about the face and head, and had a gash across one of his ears."[17]

Pieces of wreckage were still scattered around the scene of the washout in Thaxton, but construction of a trestle had begun on Tuesday night. By six o'clock in the morning on Thursday, July 4, the Norfolk & Western road through Thaxton was open for business once again. The first train to test the new trestle was an engine towing a car with the railroad commissioner James C. Hill aboard. It was his first trip to the accident site because he had been ill when the wreck occurred.

The passage over Wolf Creek was declared safe. The 7:40 a.m. train from Lynchburg was scheduled to depart on time west through Thaxton and on to Pocahontas, Virginia. Aboard this train, designated train Number Fifteen, was Augustus Becker. Becker had come from Richmond to Lynchburg the night before. His next stop was Roanoke to speak with railroad officials about the search for his friend and employee, Harry Wheeler. As the train passed over Wolf Creek, Becker could see a "few charred timbers and the iron-work of the cars and engine which had been twisted and turned into a shapeless mass by the heat of the fire."[18]

He arrived in Roanoke and met with several Norfolk & Western officials. He could not say with certainty that Harry Wheeler had actually been on the train, and he needed to get confirmation one way or the other. Wheeler had an annual pass for the railroad because he traveled so much. Becker asked the railroad officials to send out a description of Wheeler and his pass, number 914. A telegram was sent off to the conductors who managed passenger train Number Two before it reached Roanoke. There were two conductors, Stanley and Blanchard, who were responsible for the runs from Bristol to Radford, Virginia, and from Radford to Roanoke. Both conductors replied with news that further diminished Becker's hope that Wheeler was not on the train. Conductor Stanley replied, "Yes, sir. A man filling your description was on No. 2 Monday night last." Blanchard confirmed, "Man with Pass No. 914 was on No. 2 Monday night."[19]

The only chance that Wheeler escaped death was the slim possibility that he decided to get off the train at Roanoke. Becker went to visit the man who would know for certain, Conductor Roland Johnson. When Becker arrived, Johnson was in bed and recuperating from his injuries. He asked Johnson if he happened to remember a Harry Wheeler on the train. At first Johnson could not recall a Harry Wheeler, but when pass 914 was mentioned, Johnson made the connection. He remembered his brief conversation with Wheeler, and his description of Wheeler's appearance convinced Augustus Becker that his friend was gone. He traveled to Thaxton to conduct his own search for some vestige of Harry B. Wheeler's remains. He was unsuccessful. Becker would have to go home to Richmond and tell Harry's wife, Laura, that she was a widow and that little Arthur would have to grow up without his father.

Becker, of course, was not the only man from Richmond searching for the remains of an employee and friend that day. John Bowers met his son John Jr. at Thaxton to search for evidence of John Stevenson. They had no doubt that Stevenson had been in the wreck, and their only wish was to find something left of the man so that they could carry it back to Richmond. They searched in vain. All remnants of the life of

John Stevenson had been burned in the fire and washed away by Wolf Creek. Another sad telegram traveled the wires to Stevenson's brother back in Richmond. His sorrowful task would be to tell Ada Stevenson she too had been widowed in the disaster at Thaxton.

As if death and loss alone were not enough to torment those left behind after the wreck, the fire that had consumed what was left of those who perished added to the grief. After spending an entire day searching for anything that might be left of their sons, the fathers of John Hardwick and Will Marshall headed home on the morning of July 4. They had failed. The anguish of the fathers was palpable even in the telegram sent home by Christopher Hardwick to announce the terrible news. "Our last hopes gone. Have been to the wreck and find no trace of the boys."[20]

Like most of those killed in the wreck, Marshall and Hardwick's remains would never be found. The flames had denied these families the opportunity to give their loved ones a final resting place. The patriarchs of the Marshall and Hardwick families returned to Cleveland, but it was not the homecoming they had hoped for.

As the disconsolate fathers headed home with empty hands and hearts, the friends and family of John Kirkpatrick were in Roanoke and desired to pay their respects to Kirkpatrick. They had been searching at Thaxton alongside Marshall and Hardwick's fathers and had found no sign of John. At the undertaker's office in Roanoke, two coffins contained the small quantity of human remains that had been salvaged from the wreck. The coffins contained at least six spinal columns, and one small heart believed to be that of baby Charlene Peyton. No identifiable trace of her mother, Jessie, or her father, Charles, had been found. John Kirkpatrick's loved ones felt that if any part of the young man was possibly in those coffins, they wanted to attend the burial.

On the morning of July 4, Kirkpatrick's friends and family followed behind the two coffins as the undertaker, George Sisler, carried the remains to the Roanoke City Cemetery. One large grave stood ready to receive the coffins. Reverends William Creighton Campbell and John Bushnell conducted a small graveside service that was sparsely attended because most of the people in town had no idea it was taking place. No notice had been published about the burial. The tears of Kirkpatrick's friends and family were mostly all that were shed at the city cemetery that morning.

The coffins were placed side-by-side in that single grave, and flowers were placed carefully upon the fresh mound of dirt. Some would later accuse the railroad of hurriedly burying the remains to conceal the number of lives lost. No matter the reason, the unannounced burial was a tragic injustice to those who had died so terribly at Thaxton.

Some newspapers around the country continued to publish news coverage of the wreck for several days afterward. One of these newspapers came to the attention of Mr. George Rutledge Stuart, a Methodist minister from Cleveland, Tennessee. George was in Buffalo, New York, on his way home after a trip to Canada. Although he had enjoyed his time boating across Lake Ontario with his friend Sam Jones, it was not the trip he had originally been dreaming about.

The trio of friends from Cleveland on passenger train Number Two had initially been a quartet. George had been planning the trip with Will Marshall, John Hardwick, and Will Steed for months. Unlike the single fellows, he was in his early thirties, married, and had two children already. His father had passed away fourteen years earlier, and his mother, Maria, lived with him. Despite an initial protest against the trip from his mother, he had gone ahead with all of his arrangements and thought it would not be a problem. He planned for his father-in-law to stay with the family and help take care of the children while he was gone.

The time finally arrived to leave, and George told his mother that he was off for Europe soon. George described the conversation that followed. She told him, "George, I told you once I did not want you to go. I have thought over this trip and prayed over it, and I cannot give my consent for you to go; and now I tell you so that you will understand it: You shall not go."[21] He pleaded with her to change her mind and told her, "It is one of the sweetest hopes of my life that you are crushing."[22] She would not change her mind, and he gave up the fight. The battle was lost, and he had to face his friends to deliver the news.

Steed, Hardwick, and Marshall considered George's situation the same way many men would. George was a grown man, married with children, and should not be tied to his mother's apron strings. He explained his unshakeable loyalty to his mother and declared, "I would not cross the old Atlantic against my mother's wishes for a million dollars."[23] With that clearly stated, it was official. Steed, Hardwick, and Marshall headed to Europe without George.

A few days later, he received an invitation to accompany his friend Sam to Canada. As they headed to a dining room for supper in Buffalo, his friend noticed the headlines on one of the New York papers. "George, there has been a terrible railroad wreck at Thaxton, Va. My! What a list of the killed!" Sam exclaimed.[24] George glanced at the paper and saw his own hometown listed next to some of the names. His blood ran cold as he read the names of his three friends, Hardwick, Marshall, and Steed among those killed. "Oh, Sam, the next name would have been 'George R. Stuart, Cleveland, Tenn., killed and burned,' but for the authority of my precious mother!"[25] George returned to Cleveland immediately and found the town gathered in the streets in mourning. He saw the mother of one of the three who were killed. She exclaimed, "O George, if I only had the body of my precious boy!"[26] When he saw his own mother, she ran to meet him at the gate, and she cried out, "Thank God! My boy is safe."[27] Honoring his mother's wishes had saved George Stuart's life.

Stuart returned to Cleveland and spoke at a memorial service for his friends on Sunday, July 7. The day became an unofficial day of remembrance for the victims of the wreck at Thaxton. Memorial services were held across the state of Virginia and in Tennessee. Stuart spoke at an afternoon service conducted at the Methodist church in Cleveland, one of two memorial services held in the city that day. In Richmond, a tearful service was held in the morning at the Presbyterian church for little Pattie Carrington. In Roanoke, people had not been informed of the graveside services that had taken place on Thursday, but they gathered and mourned as a city Sunday evening. A large congregation attended the Presbyterian church there in the city that had lost more of its citizens than any other.

A majority of the injured had made it home from the hotels in Roanoke within a week of the wreck. Bishop Alpheus Wilson had arrived at his home in Baltimore on July 4. The splinter from the wreck that had run through his palm still had to be removed. His doctor suggested chloroform as an anesthetic during the painful removal process. Alpheus declined. He insisted, "If I can have a good cigar, I'll not be a trouble to the doctor."[28] Wilson's ability to write would never be the same again, but considering the circumstances, things turned out well for him. He was alive after a night of horror and pain lying on the tracks in the darkness and rain.

By July 8, Norfolk & Western had cleaned up all the debris from the wreck and taken it to Roanoke. While the company completed the salvage work, workers had found two more bodies burned among the ashes. The official number of deaths reported by the railroad was seventeen, but that number would later prove to be inaccurate. Of the seventeen the railroad identified as killed, twelve of them were listed as passengers on the first-class coach. Only four men riding on that coach survived. J. A. Young, Fred Dexter, Robert Goodfellow, and Fred Temple miraculously had survived with mostly minor injuries.

The true count of those killed would never be known for sure. At least eighteen lives were taken, and twenty-one others were injured. Fifteen passengers and three crew members were gone. The entire Peyton family was erased in an instant. Two little girls would never again be able to shine the joy of childhood on those left behind. While the forces of nature had conspired to create this terrible event, it is human nature that causes us to forget easily what was lost. At the time of the wreck, it was the worst accident in the history of Virginia railroads. On July 8, the last official bare-knuckled fight in the country's history between John Sullivan and Jake Kilrain went seventy-five rounds and forced nearly every other story off the front page of most national newspapers. A few of the Virginia newspapers printed small memorials for some of those who were killed. In Cleveland, several young men were already working to have a monument in town constructed to honor the three men they had lost. The city had loved Steed, Hardwick, and Marshall so much that over $1,000 was received within two days of starting up a collection for the monument. It arrived in Cleveland by late November the same year, and the monument remains in the center of town to this day.

In stark contrast, at Roanoke, a hastily dug common grave contained the remains of an unknown number of victims. Over the years, the burial site was lost. A rebuilt fill over Wolf Creek remains at Thaxton, and no marker or monument to recognize the souls lost at the site had been erected prior to the release of this book. Railroad accidents with much smaller death tolls were frequently marked in some way, and in some cases even celebrated in popular songs. Somehow, the tragedy at Thaxton was forgotten.

Alpheus Waters Wilson had always carried a Greek Testament with him wherever he traveled. That Testament was one of many personal items lost in the wreck along with the lives taken. Wilson had carried it for nearly twenty years and read four chapters from it daily. Along with it, he had lost his watch and many other items he used every day. In the years after the wreck, he would frequently mention that he needed something that had been "lost at Thaxton."[29]

Personal items could be replaced, but the lives destroyed that night at Thaxton were lost to eternity. This book will not bring them back, but their memory and their story are no longer lost.

The wreck at Thaxton, viewed from the southern side of the washout. (Photograph courtesy of Norfolk and Western Historical Photograph Collection, Norfolk Southern Archives, Norfolk, Virginia. Digital image courtesy of Special Collections, Virginia Tech, Blacksburg, Virginia.)

The wreck at Thaxton, viewed from the northern side of the washout. (Photograph courtesy of Norfolk and Western Historical Photograph Collection, Norfolk Southern Archives, Norfolk, Virginia. Digital image courtesy of Special Collections, Virginia Tech, Blacksburg, Virginia.)

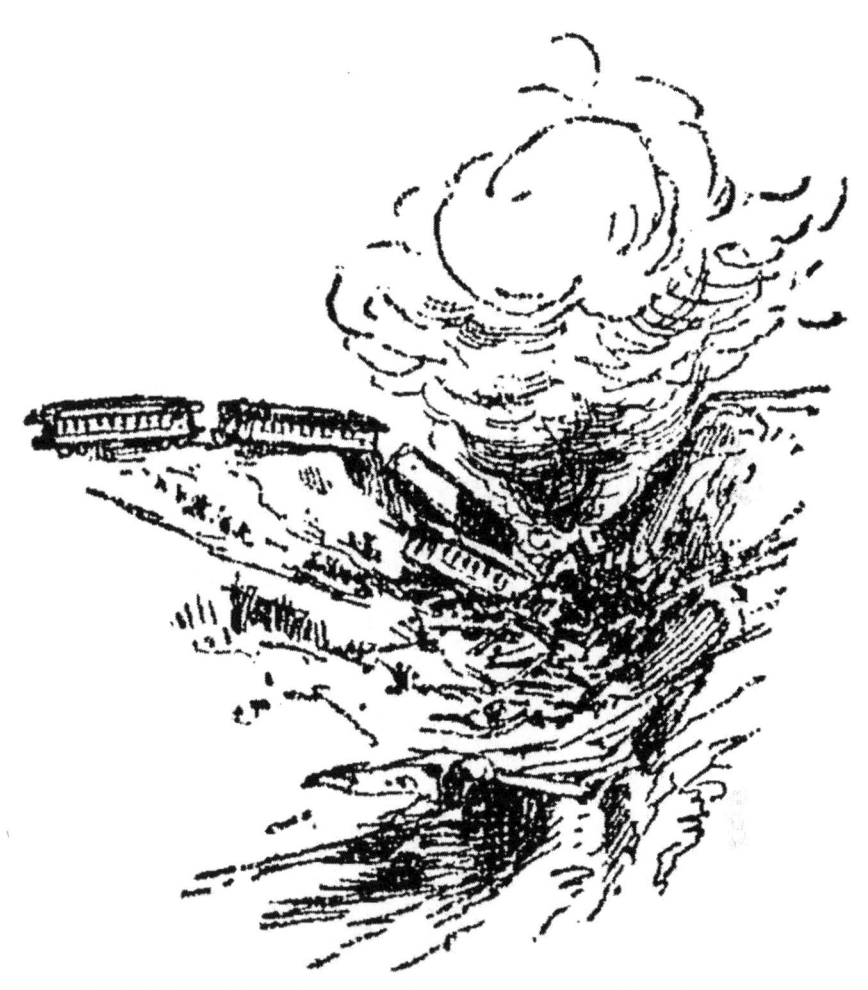

Sketch of the wreck drawn by passenger George Masters. Reprinted from "The Virginia Railroad Accident," Evening Repository (Canton and Massillon, Ohio), July 9, 1889.

Three of the doctors who treated the wounded at Thaxton. From left to right: Samuel Price (Buford, Virginia), Joseph Gale (Roanoke, Virginia), Herman Jones (Roanoke, Virginia) Reproduced from Lyon G. Tyler, ed., Men of Mark in Virginia (Washington, DC: Men of Mark Publishing Company, 1908), 340–342.; H. A. Royster, ed., Transactions of the Southern Surgical Association (White Sulphur Springs, WV: Southern Surgical Association, 1917), 618–620.; George S. Jack and Edward Boyle Jacobs, History of Roanoke County (Roanoke, VA: Stone, 1912), 201.

Because the city of Roanoke did not yet have a hospital, most of the injured from the wreck were taken to Hotel Roanoke. (Photograph courtesy of Norfolk and Western Historical Photograph Collection, Norfolk Southern Archives, Norfolk, Virginia. Digital image courtesy of Special Collections, Virginia Tech, Blacksburg, Virginia.)

Monument placed in Cleveland, Tennessee, to memorialize the three Cleveland men killed in the wreck at Thaxton. The photo is undated but was likely taken prior to 1911, when a Confederate Monument was erected next to it. The monument stands in downtown Cleveland to this day. (Courtesy of the History Branch and Archives of Cleveland Bradley County Library)

Engine Number 30 bell clapper recovered from the wreckage. (Courtesy of Bedford Museum and Genealogical Library)

Culvert at Wolf Creek. The culvert is now much larger than the three-foot by four-and-a-half-foot culvert that was unable to carry the water away in 1889. (Taken by author in April 2012)

*Culvert and fill over Wolf Creek. The railroad still passes over at the top of the fill.
(Taken by author in April 2012)*

Former location of the Thaxton Station. The station's right-hand side was located approximately where the stack of railroad ties is seen. (Taken by author in April 2012)

Looking west from the location of the Thaxton station. Note the milepost on the right marked with the number 234. This was the same milepost number used in 1889 at the location of the station. (Taken by author in April 2012)

The Peaks of Otter rising above Thaxton. (Taken by author in April 2012)

PART II

THE REST OF THE STORY

8

BLAME

"We believe, therefore, that the officials of this road exercised all care and prudence in the running of this train that could reasonably be required of them, and that the company is not to blame for this terrible and lamentable disaster."[1]

— BEDFORD COUNTY GRAND JURY, JULY TERM, 1889

When something as horrific as the wreck at Thaxton occurs, it is human nature to look for someone or something to blame. In some ways, laying blame helps to eliminate feelings of helplessness that tend to follow in the wake of such a tragedy. People also have a deep-seated need to see some sort of justice done and restore order and balance to the world.

Not long after the wreck, the wheels of justice were set in motion. Although this accident was like no other the state of Virginia had ever experienced, railroad accidents in general were common. Numerous lawsuits to recover damages to life and property were always the inevitable result. Norfolk & Western moved quickly to deal with the legal ramifications it faced after the wreck. The company employed some time-tested corporate measures right away to pressure William Ford, one of its own employees.

Ford was the baggage master on the train and had sustained some significant injuries in the wreck. In addition to scald burns, he had sustained head and spine injuries. By July 17, 1889, just two weeks after the wreck, word had somehow reached the N&W corporate offices that Ford was considering a lawsuit against the company. Division Superintendent James Cassell had been in the baggage car with Ford when the train went down. Cassell composed a letter to Ford in which he dangled a carrot and a thinly veiled threat concerning any plans Ford had to file a lawsuit.

In the letter, he inquired about Ford's health and went on to mention a promotion they had previously discussed before the accident occurred. The job was the night ticket agent's position at Roanoke. Any role that would get him off the rails on a regular basis would be an attractive one, and after the incident at Thaxton, it likely sounded better than ever to Ford. Cassell went on in his letter to mention a discussion he had with General Manager Sands to make the position an even more desirable daylight job.

After the opening pleasantries and talk of promotions, Cassell then addressed any thoughts of a lawsuit that Ford might be considering. He referred to the potential lawsuit and addressed it by saying, "If this is the case, of course, it is useless for me to take any farther steps in connection with the agency matter."[2] In other words, if you file a lawsuit, don't even think about a promotion. Cassell hammered the point home by finishing the letter with, "I would just say, for your information, that the company admit no liability on account of the accident at Thaxton and in my opinion, they should not."[3] The message was loud and clear. If Ford filed his lawsuit, he could say goodbye to his career with Norfolk & Western and the company intended to fight the case tooth and nail.

Nearly two weeks passed after Cassell wrote his first letter to Ford, and there was no reply. Cassell requested an update in a follow-up letter on July 29. He dangled the promotion opportunity again and indicated that he was "about to make arrangements for permanently filling the vacancy in Ticket Office at Roanoke."[4] The promotion was not the only consideration for Ford. In the same letter Cassell let Ford know that his current job was at risk by clarifying, "I also desire to know what action to take in regard to filling your place on the train."[5] It may have been the simple need for the company to fill a necessary role for its business, or it could have been a heavy-handed tactic in case Ford was still thinking about a lawsuit.

Cassell and N&W's confidence in their position was boosted the very day his follow-up letter was drafted. A Bedford County grand jury had been assembled to review the events of the accident and determine whether the company was at fault. The grand jury consisted of eight men who were primarily businessmen and farmers. They concluded that Norfolk & Western was not to blame for the accident. One of the key pieces of information that led the grand jury to its conclusion was the length of time the train had taken to cover the distance from Buford to Thaxton. In general, a fast train would be able to cover the twelve miles from Buford all the way to Liberty in about fifteen minutes. The night the accident occurred it had taken over twenty minutes just to get from Buford to Thaxton, and that was still five miles shorter than the full trip to Liberty. If the railroad had been negligently running the train too fast for the stormy conditions, the trip would not have taken that long. Experts had also testified that they believed that when the train arrived at the fill, the track was still intact. This testimony would, of course, debunk any theory that a fast-moving train could have leaped into the washout. Although these were the details that persuaded the jury to exonerate the railroad, members of the jury did point out that no one could know for sure. The full text of the jury's decision is shown here:

> *To the Hon M. Davis, Judge of the County Court of Bedford County:*
> We, the undersigned members of the grand jury empanneled for the July term of the county court of Bedford county, who were specially charged to investigate the recent disaster on the Norfolk & Western railroad, above Thaxton, in this county, respectfully report that after a careful examination, and investigation into all the facts in connection with the disaster, we do not find that any blame can be attached to the Norfolk & Western railroad company, or any of the officers or employees.

We carefully investigated the condition of the culvert where the washout took place, and we find that it was in good condition, that it has been standing ever since the road was built, about thirty-seven years, and that it has been examined, as well as other portions of the road, every year for the last several years, by experts, and reports made on it. These reports show that it was in good and safe condition; and in fact the proof shows that no part of the road was in better condition than that between Bufordville and Liberty, half way between these two points to where the disaster occurred. The distance between Bufordville and Liberty is twelve miles, and the fast train usually runs it in fifteen minutes, but on the night of the accident it was twenty-five minutes in running from Bufordville to the culvert, which shows that this fast train was running at an unusually slow rate of speed. This precaution was adapted in consequence of information gotten at Bufordville, that there was indication of unusually heavy rainfall below. The rain was light at Bufordville, in fact the heavy rain extended but a very short distance west of the culvert; but from that point several miles down the mountain the rainfall was not only unusually heavy, but heavier, very much more so, than any that could be recalled in the memory of the oldest inhabitant; it was in fact a perfect water spout or cloud burst. The washouts, and land slides caused by that rain (on the sides of the mountains) are visible for long distances.

While no one can speak with certainty on this point, yet from testimony by experts, the track above the culvert must have been intact when the train came upon it. We believe therefore, that the officials of the road exercised all the care and prudence in running this train that could be reasonably required of them, and that the company is not to blame for this terrible, and lamentable disaster.

Signed,
R.B. Claytor,
J.W. Gish,
Samuel R. Mead,
W.P. Hoffman,
John W. Lowry,
John W. Johnson,
T.D. Berry,

J.E. Nelms.

— *(REPORT OF THE RAILROAD COMMISSIONER, 1890, P. XLI–XLII.)*

In the grand jury's opinion at least, blame for the accident would fall under the category of "Act of God." Since a lawsuit against the Lord would be neither feasible nor advisable, the grand jury's findings were not enough to prevent others from suing Norfolk & Western.

Within a day or two after newspapers across the country published the grand jury findings, William Ford's wife, Mary, finally composed a reply to James Cassell. Mrs. Ford was very concerned about her husband's health, even though Dr. Walter Izard had told her he was in no danger. Izard was one of the Liberty doctors who had gone to the scene of the wreck. Mary Ford was not particularly confident in his abilities. In her letter to Cassell, Mary related a story about a sick child of hers who was given a clean bill of health by Izard, only to have the child die four hours later. She believed her husband was in much worse condition than Dr. Izard had diagnosed.

Mrs. Ford was also noticeably disappointed with Norfolk & Western's management. She described the promotional opportunity offered by General Manager Joseph Sands as "too late to offer 'Justice'" and pointed out that her husband was "in no condition to either mentally or physically to consider it."[6] Her anger seemed directed more at Joseph Sands than Cassell. She thanked Cassell for aid that he had rendered to her husband while they were at the wreck site. She finished her note to Cassell with a request that they send one of the Roanoke doctors along with Izard the next time they came to check on him.

Cassell complied with Mrs. Ford's request. He dispatched Dr. Arthur Koiner from Roanoke to visit the baggage master and his wife at their home in Lowry, just outside of Liberty. Koiner was another of the doctors who treated the wounded at Thaxton. Mrs. Ford did not find an ally in Dr. Koiner. In his opinion, her husband's condition was a result of his refusal to follow the advice of Dr. Izard and that he was not taking proper care of himself. Cassell passed on this information to

Mrs. Ford in another letter and once again inquired about her husband's possible intention of filing a lawsuit against the railroad.

William Ford did end up filing suit against the railroad along with at least seventeen others. Most of the cases were filed within a year of the wreck. Some like Will Marshall and Patrick Donovan's families sued because they had lost their loved ones, while others sued because of injuries they sustained in the wreck. The larger cases, like Ford's, were generally seeking $10,000 in damages, and many of the injury related cases were filed seeking $4,000. The same two attorneys, James Bible and Frank Blair, represented a majority of the plaintiffs. Bible was the man who traveled to Roanoke after the wreck and escorted Will Steed's body home to Cleveland. Blair was a former state attorney general for Virginia.

Norfolk & Western countered with a few different attorneys, including Colonel Abe Fulkerson from Bristol. Fulkerson had been a colonel for the Confederate army and was well known as one of "The Immortal 600" during the war. The Immortal 600 was a group of Confederate officers who were prisoners of war used by the Union Army as human shields on Morris Island in Charleston, South Carolina. The Union kept them there for six weeks while being shelled daily by both Union and Confederate guns, and Fulkerson emerged as one of the leaders for the tortured prisoners. He was not a man likely to fear sparring in the courtroom.

Fulkerson's six weeks of endurance during the war were quite brief compared to the years N&W battled with those who felt the company was negligent the night of the wreck. Norfolk & Western seemed to be on the winning track in all the cases until June of 1893. A jury awarded $7,500 to the family of Will Marshall, one of the three men from Cleveland, and Marshall's family became the first to win a suit against N&W for the wreck at Thaxton. They could not celebrate the victory for long, however, because the Supreme Court of Appeals of Virginia reversed the decision eighteen months later. The appeals court based its decision on the concession that an act of God had caused the accident and that there was no proof N&W had been negligent. Once again, the courts pointed the finger at the Almighty.

While the official effort to assign blame took place in courtrooms, the court of public opinion was also at work. The Bedford County Grand Jury had found nature at fault, and longtime residents in Bedford County agreed that the storm the night of the wreck was unlike anything they had ever seen. Others felt that a big storm was not reason enough to let the railroad off the hook. The thinking was that nature will always be unpredictable, but the railroad companies should do a better job of providing safeguards and spending the money it would take to provide those safeguards.

The wreck at Thaxton briefly shone a spotlight on issues that the nation's railway system needed to address. Some called for more government oversight of the railroad companies. The *Daily Picayune* in New Orleans suggested new Federal laws for inspection of trains, roadbeds, bridges, and requirements for the qualifications of railroad employees.

Virginia created the office of Railroad Commissioner about twelve years prior to the wreck to provide for oversight and regulation on a state level. The commissioner at the time of the wreck, James Hill, had visited Thaxton to form his opinion on what happened that night. His judgment was that "he failed to discover wherein the accident could be attributed to any negligence or carelessness on the part of the railroad authorities."[7] The train's slow progression from Buford to Thaxton, the extra man riding on the engine, the long-standing safety record of the embankment, and the other weather-related destruction in the area all contributed to Hill's conclusion that the railroad company was not to blame.

There was a problem with Hill's assessment and the idea in general that what happened at Thaxton was simply a freakish natural disaster. If it were looked upon as "just one of those things," and, with a shrug of the shoulders, everyone went back to business as usual, then lives would continue to be lost without true analysis of how to make the system better. An attorney in Richmond questioned Hill's conclusion that there was no one to blame for the wreck. In a letter to the editor of the *Richmond Dispatch* on July 9, 1889, he asked, "Is it possible that such a sacrifice of human life has taken place and there is no one to blame? I am not one of those who would lay such things to the door of accident."[8] The attorney went on to discuss a conversation he had with an official from the Pennsylvania Railroad about how that well-respected company managed its tracks. The practice of the Pennsylvania Railroad was to have a track-walker examine track before the passage of every train. The official theorized that Norfolk & Western did not follow the same procedure in order to save money. The attorney's position in his letter to the editor was that the wreck should be considered "manslaughter" because N&W did not have a trackwalker inspecting the track ahead of passenger train Number Two.[9] In fact, N&W did have watchman John Johnson assigned to duty that night, but it was not widely known that he had been trapped in his own house by another washout on the other side of the station.

Manslaughter was too weak a charge for the *Free Lance* newspaper in Fredericksburg, Virginia. In the July 9 edition, the paper expressed its opinion that the wreck was a case of murder. "*We call it murder because it was murder* and not 'accident' as the friends of that road proclaim it was, and until some of those responsible for the care and protection of human lives are hung, such murders will continue."[10] The paper went on to cast blame specifically on a system in which "as few men as possible are employed and paid to watch the tracks, examine the machinery, and in other ways, provide against these so-called accidents."[11] Norfolk & Western's move to prevent the press from visiting the scene also stirred up suspicion. It caused the *Free Lance* to doubt reports that a track-walker had been on the job that night. The paper charged that N&W "gathered as soon as possible from the burned cars every vestige of human remains that could be found and boxed and buried them to prevent identification."[12]

While the *Free Lance* could certainly have been accused of a little exaggeration, the point was a valid one. Nothing could be done to prevent the power and unpredictability of nature, but the railroad companies could certainly change how they dealt with those situations. Even if it meant the bottom line would have to suffer due to additional labor costs, the preservation of human life was more than worth it. A quote printed in the *Richmond Dispatch* on July 9 summarized this point.

"The laws of all the States should require every mile of railroad operated at night to be strictly patrolled. The track-walker should pace his beat with the vigilance and regularity of a sentinel on guard in the face of an enemy."[13]

While many could agree that additional personnel might have helped prevent the type of accident that occurred at Thaxton, some felt that the railroad was more directly negligent than that. A number of newspapers reported that the train was traveling at a high rate of speed when it hit the washout at Thaxton. If that were the case, then Norfolk & Western would certainly have been reckless in its operation of the train given the conditions that night. The speed of the train was, therefore, a key detail in the discussion of blame for the disaster.

N&W maintained that the train was moving slowly and cautiously and that the saturated fill gave way only after the weight of the train caused it to collapse. Many newspapers described the wreck quite differently and reported the train was going anywhere from thirty to forty miles per hour, leaped across an already washed out fill, and slammed into the other side.

One of the primary sources of information the newspapers used as a basis for the fast-moving train theory was the account of Bishop Alpheus Wilson. His description of the wreck printed in the July 6 edition of Lynchburg's *Daily Virginian* newspaper stated that the train was going forty miles per hour when it fell into the washout. The *Free Lance* newspaper in Fredericksburg specifically cited Bishop Wilson's narrative in an article titled, "The Norfolk and Western Railroad Holocaust."[14] There was one fundamental problem with the Bishop's statement however. In the same account he had given concerning the speed of the train, he also mentioned that he was asleep when the train wrecked. Wilson's experience at Thaxton was retold at a memorial service for him years later, and the story remained consistent. He was asleep and woke up only once the train fell into the washout. If he were asleep before the wreck occurred, he would not have had a very good idea of the speed of the train while he slept.

Wilson was not the only one on the train who suggested it was traveling too fast for conditions. Lewis Summers, the postal clerk, told the

Daily Virginian on July 4, "It was his impression that the train was running at a very high rate of speed" when the wreck occurred.[15] Unfortunately, his story suffered from the same flaw that Bishop Wilson's did. In that same article, Summers mentioned that he had gone to sleep shortly after the train had passed Blue Ridge, and he told the newspaper that he had probably only been asleep for two or three minutes before the crash. In fact, he had been asleep quite a bit longer.

Summers testified in the civil case brought against N&W by the family of John Hardwick a year after the wreck. In his testimony, Summers revealed that he had fallen asleep shortly after they left Blue Ridge. He was sleeping so hard that he was not even aware that the train had made a stop at Buford after that. He described himself as being "rather in an unconscious condition" at the moment the train fell into Wolf Creek.[16] Under questioning from "Immortal" attorney Abe Fulkerson, Summers acknowledged that he had no idea how fast the train might have been traveling after he went to sleep. Fulkerson now was the one doing the shelling and shooting holes through the fast-moving train theory.

There was one man awake on the train that felt it was traveling more quickly than it should have been, Baggage Master William Ford. Ford testified in Bedford County Circuit Court in 1893 during the case brought by the family of another of the Cleveland, Tennessee, men, Will Marshall. Ford's opinion was that the train was moving about thirty to thirty-five miles per hour. He based his opinion on the sound of the wheels as they passed over the tracks and the general motion of the car as they traveled that night. As a twenty-five-year veteran of the railroad, Ford would certainly have a refined feel for how fast a train was traveling. There was one conspicuous issue with Ford's testimony. He had already filed his own lawsuit against the railroad seeking $10,000 in damages. It would certainly have been to his own benefit to suggest that Norfolk & Western had not managed the trip cautiously that night.

So what was the truth concerning the speed of the train when it arrived at Wolf Creek? There were two competing theories to consider. One was the description put forth by Norfolk & Western that the train was on the fill when the earth gave way underneath it. The second was that the washout was already there when the train arrived, and it was traveling at such a speed that it launched across the eighty-foot-wide gap and embedded itself into the eastern side of Newman's Fill.

The Bedford County Grand Jury was presented with testimony from experts and referenced that testimony when it concluded that "the track above the culvert must have been intact when the train came upon it."[17] The time the train left Buford and arrived at the culvert was well documented, and twenty-five minutes had passed just to cover six and a half miles. That adds up to an average of around fifteen miles per hour for that leg of the trip.

The location of railroad employees on the train such as Alvin James and James Cassell also gave some insight into some precautionary measures they had taken. James was only a passenger that night but had moved to the engine to ensure that they proceeded carefully when they left Buford. Cassell's location in the baggage car indicated as well that they were at the very least attempting to keep a watchful eye on things.

Further evidence of a slow-moving train came from Stephen Hurt, a track watchman. He had described carrying on a conversation with Engineer Donovan while the train slowly rolled by his position. That was at a point only two to three miles from Wolf Creek.

It is possible that once the known trouble spots were cleared that Patrick Donovan decided to push the throttle hard the last two miles before they were to arrive at the Thaxton station. There were at least two newspaper reports that mentioned the throttle position of the wrecked train. The *Daily Times* in Richmond, Virginia, printed an article that stated the throttle was in a position that indicated Donovan "had his engine going with great rapidity of speed."[18] Another Richmond newspaper, the *Richmond Dispatch*, also reported that the throttle position demonstrated Donovan "had his engine wide open and was running at a rapid rate of speed."[19] Newspapers were not always completely reliable, however, especially when it might get in the way of telling a good story that would sell papers. If there were not enough drama in the truth, then writers used a little embellishment from time to time. For example, the paper might have an employee climb into a tree and pretend to be a corpse, creating an intriguing photo of a poor soul who had been deposited there by some great flood or storm.

If Donovan had been pushing the train hard and it had leaped across a chasm, would the passengers and crew on Calmar have described the wreck as they did? Those in the last car described their experience as simply a "sudden jerk" or something similar to the train applying air brakes.[20] Samuel Boyd indicated that he was not even knocked from his stool on the rear platform of Calmar. When the train wrecked, J. H. Elam was standing up in Calmar and was not thrown off his feet. If the engine was traveling at such a rate of speed that it could make an eighty-foot jump, it seems the momentum would have affected Calmar more than it did.

Burton Marye was a passenger on the train, and in his opinion the position of the cars after the wreck suggested that the fill was intact and that "The engine did not make a plunge as was stated, but the cars sank rapidly as the bank was melted beneath them."[21] The strength of his assessment was bolstered by the fact that he was a civil engineer, and he had worked around the railroad throughout his career. Even a correspondent for the *Daily Virginian* newspaper who had made it to Thaxton the day of the wreck agreed with Marye's theory. In his report printed on July 3, the correspondent suggested the fill gave way as the train passed over it.

Despite reasonable evidence that the train did not make a leap, there are certainly valid arguments that could be presented against that evidence. The grand jury's opinion could have been biased. It was composed of several prominent Liberty businessmen who could have been inclined to lean the railroad's way in the name of preserving commercial relationships. There should be no assumption that this group was a beacon of truth, honor, and integrity. At least two of them, R. B. Claytor and T. D. Berry, were later indicted for aiding and abetting embezzlement and swearing to false bank statements as executives at the First National Bank of Bedford City.

Employees like Elam, Boyd, Hurt, and others who gave statements relating the train's slow speed might have been unwilling to say anything that would have put the company in a bad light. Norfolk &

Western demonstrated they were willing to play hardball with employees with their handling of William Ford. Even Burton Marye's opinion might be tilted, even though he was not an N&W employee. He was still a railroad man.

The strongest evidence points toward the "fill gave way while the train was on it" instead of the possibility that the train leaped across the washout. That evidence includes the grand jury's analysis, the amount of time that passed between stops, the position of the cars in the wreckage including Calmar, the descriptions by passengers and crew on Calmar, Burton Marye's opinion, and the railroad commissioner's opinion. These details seem to lead away from the idea that the engine took a high-speed leap across an open washout.

Only three men knew for sure exactly how fast the train was going when it arrived at Wolf Creek, but they all perished when the engine tumbled into the washout.

9

WHO IS JOHN DOWELL?

"You will please find a full and complete report of the disaster which occurred about one-half mile west of Thaxton, on the morning of July 2d, 1889, in which passenger train No. 2, east-bound, was first wrecked and then destroyed by fire, and by which seventeen persons were killed and twenty-one injured."[1]

— JOSEPH H. SANDS, NORFOLK & WESTERN GENERAL MANAGER, LETTER TO N&W PRESIDENT FREDERICK KIMBALL, JULY 29, 1889

In the weeks after the wreck, Norfolk & Western compiled a detailed record of the events that transpired on that terrible night at Thaxton. In 1890, the report was included in the *Fourteenth Annual Report of the Railroad Commissioner of the State of Virginia*. The "official" list of those who were killed in the wreck at Thaxton was included in the report. A copy of the list appears as follows:

LIST OF PERSONS KILLED IN THE WRECK.

NAME.	OCCUPATION.	ADDRESS.	LOCATION ON TRAIN.
A. M. James	R. F. of E.	Roanoke, Va.	Engine No. 30
Pat Donovan	Engineman	Lynchburg, Va.	Engine No. 30
J. E. Bruce	Fireman	Roanoke, Va.	Engine No. 30
J. J. Rose	Mail clerk	Abingdon, Va.	Postal car.
J. W. Lifsey	Train Desp'r	Roanoke, Va.	Coach No. 63
Dennis Mallon	Janitor	Roanoke, Va.	" " "
John Kirkpatrick	Passenger	Lynchburg, Va.	" " "
Nathan Cohn	"	Roanoke, Va.	" " "
Chas. L. Peyton	Stenographer	Radford, Va.	" " "
Mrs. Chas. L. Peyton	Passenger	Radford, Va.	" " "
Chas. Peyton, (child)	"	Radford, Va.	" " "
W. C. Stead	"	Cleveland, Tenn.	" " "
J. M. Hardwick	"	Cleveland, Tenn.	" " "
Wm. F. Marshall	"	Cleveland, Tenn.	" " "
H. B. Wheller	"	Chattanooga, Tenn.	" " "
J. I. Stevenson	"	Richmond, Va.	" " "
Miss Patty Carrington (child)	"	Staunton, Va.	Tobocco.

Norfolk & Western's report included some misspellings, inaccurate hometowns, and an ambiguous description of baby Charlene Peyton, but the names in the report were easily reconciled with the numerous newspaper articles in 1889. The railroad had nearly a month to compile the list, and the area newspapers in 1889 had not revealed any new names. Surely if there were others, the newspapers would have been the first to break the news, especially considering their general disdain for the way Norfolk & Western had dealt with the press just after the wreck. Some newspapers printed early "estimates" of the dead at some number higher than seventeen, but there were never any names attached to validate any additional deaths beyond those listed in the railroad's report. The numbers presented by General Manager Sands in the report, seventeen dead and twenty-one injured, appeared to be full and complete as Sands had described.

The list was not complete.

A small file in the archives of the Bedford County Courthouse contains a record for the case of *Dowell's Administrator versus Norfolk & Western Railroad Company*. A man named Charles M. Harper filed the case in July of 1890, one year after the wreck. Harper charged that John T. Dowell was killed in the wreck at Thaxton and that he was the administrator for Dowell's estate. Harper was seeking $10,000 from Norfolk & Western on behalf of Dowell's estate, claiming the company had been reckless and careless in their operation of the train. The case file included a note indicating that Norfolk & Western's attorneys "craved oyer" for Harper's documentation that he was actually the administrator for Dowell's estate.[2] Apparently, Harper was not able to provide the documentation or what he did provide was not sufficient. By the end of the year after Harper filed the case, a judgment was made to reimburse Norfolk & Western for its court costs.

Dowell's name never appeared in the major newspaper articles that covered the wreck in 1889. Unfortunately, the court records also provide no additional clues to Dowell's identity. There is no testimony in the case file or any other information that might explain how Dowell was connected to the wreck of passenger train Number Two. Just who was the John T. Dowell that Harper claimed was killed at Thaxton?

The author of an article printed in the *Augusta Chronicle* on August 22, 1889, did not wonder who John T. Dowell was, but rather, where was he? The article described Dowell as a well-to-do farmer from the city of Rome in Floyd County, Georgia. He had been in Georgia for about sixteen years after moving from Virginia with his wife, Lavinia. She passed away a few years before 1889. John had decided it was time to head back home to Virginia, and his friends in Georgia believed he had been on passenger train Number Two when it wrecked, but they could not be absolutely sure. He had not been seen since July 2, the day of the wreck. One of Dowell's friends was quoted in the article: "All the baggage was burned in the wreck and I am absolutely without a clue as to positive proof of Mr. Dowell's death. Yet having heard nothing from him since that date, and feeling quite sure that he was on the ill-fated train I believe that he perished in the wreck."[3] The writer of the article went on to reference the fact that some hearts were recovered from the wreck, but "the trouble is that his friends cannot identify Dowell's heart."[4]

The *Augusta Chronicle* article and the Bedford County court records combine to suggest that Dowell may have been on the train, but as Dowell's friend mentioned, there was no positive proof. In the nineteenth century, it would have been easy for Dowell to lose contact with his friends in Georgia. He could have disembarked at one of the stops along his route, or he might have passed away somewhere else. Census and marriage records for John and Lavinia seem to indicate that they were married at a somewhat older age (forties), and there is no clear evidence that they had any children. John's friends would be the only ones who would have any idea where he might be, and, unfortunately, they had no hard evidence to say that Dowell was on the train the night of the wreck.

There was one small piece of information that John's friends in Georgia likely did not have. The front page of the July 6, 1889 edition of the *Richmond Dispatch* featured a large, full-page article dedicated to coverage of the wreck at Thaxton. The article, titled "Two of Our Men Lost," revealed the fates of Richmond natives John Stevenson and Harry Wheeler.[5] The paper provided other information including some additional details of the wreck and the statements of Railroad Commissioner James Hill. Like all the newspaper articles that documented the deaths at Thaxton, there was absolutely no mention of John Dowell.

Across the front page, buried about halfway down in the sixth column of an eight-column layout was a very small article titled "Gathered at Lynchburg."[6] Mixed in with snippets about the train schedule resuming normal operation, delegates appointed to the gubernatorial convention, and a woman bitten by a snake while sitting on her porch was a short paragraph that provided the final clue in the mystery of John Dowell's fate.

"A small piece of manuscript was shown your correspondent today that was picked up at the late wreck. It was partially burned, but I made out the following sentence: 'Made and entered into this 15th day of May, 1885, between J. T. Dowell of the first part and A. A. Hunt of the second.' It is believed to have belonged to one of the passengers on the wrecked train."[7]

The paper made no further mention of Dowell, and there was no mention of him in later issues. Taken by itself, that charred manuscript with Dowell's name was not particularly newsworthy. However, if Dowell's friends in Georgia had seen that paragraph, it would have confirmed for them that he was on passenger train Number Two when it fell into Wolf Creek. The friends of John Dowell would also have realized one additional troubling fact. John T. Dowell was the eighteenth person lost at Thaxton, and no one knew about it.

Why It Matters

John Dowell was killed at Thaxton, but his name was never recorded. This one man's death reveals the very real possibility that there were others that anonymously died in the wreck of passenger train Number Two. The number of deaths listed as "official" in articles reprinted in newspapers throughout the country was seventeen. For example, on July 4, the *Knoxville Journal* reported, "The names of the seventeen persons who were killed have been ascertained. The list of the killed, which is official and accurate, is as follows."[8] The article listed several names that were still inaccurate, listed Rose twice, and omitted the Peyton family, but the number seventeen remained consistent. Newspapers throughout the country printed that same list. Some articles occasionally pointed out that the number of deaths was unknown and projected much higher numbers, but there was never anything definitive. A month had already passed when Norfolk & Western provided its report for the commissioner, and its official number remained at seventeen with no mention whatsoever of John T. Dowell.

What other names were there that would never be known? Families of those who might have been lost in the flames may not have had the resources or information to file lawsuits. In some cases, they may not have realized that their loved one was even on the train. Dowell's friends at least got his name mentioned in the Georgia newspapers, but how many others did not? Ultimately, the name John T. Dowell represents an untold number whose names may never be known. Their ashes may rest unidentified to this day mixed in with the sediment and soil surrounding Wolf Creek in Thaxton, Virginia.

10

DOWN THE LINE

"After this disaster, No. 30 was in the repair shop for some time, but being thoroughly overhauled and repaired she was again put in service and is now on the fastest run on the road."[1]

— THE TIMES, RICHMOND, VIRGINIA, MAY 6, 1900

The effect of the wreck of Norfolk & Western passenger train Number Two at Thaxton was much like a rock thrown into the ocean of time. There was an initial splash at that terrible moment on July 2, 1889, and the ripples from the accident continued to spread out, even to this day. Lives ended, injuries had an impact on choices survivors would make going forward, children were left fatherless, and the mental and emotional trauma would stay with all of those who experienced it. Some would go on to tell their children and grandchildren of the wreck, and newspapers have revisited the story from time to time over the years.

Some passengers were involved in railroad wrecks again, but everyone on that train knew they had experienced something they would never forget at Thaxton. Irene Jackson, a passenger, was on her way to get married when the disaster struck. Her wedding day two weeks later may not have been the event that stayed in her memories the longest. She had not been injured in the wreck, but she summarized what most probably felt shortly after the accident. "Oh, to see the wounded—all is a picture before me continually, and I never will again travel in the night."[2] Another passenger, Fred Temple, declared, "I never passed through a more terrible experience in my life, and I assure you I have no desire to participate in such another."[3] John Scott was the Pullman Conductor on the Beverly sleeper and described the wreck as "the most horrible sight I ever witnessed."[4] Passenger Henry Martin felt the same when he said, "It was horrible, horrible, horrible."[5] For the people who were there, that awful night would never be lost in their memory.

Business As Usual

Norfolk & Western Railroad Company as it existed in 1889 would only continue for another six years. The company fell on hard times within a few years of the wreck, due in large part to an economic depression commonly referred to as The Panic of 1893. By 1895, N&W was no longer able to pay its debts, which included more than a million dollars in floating debt and $350,000 that the company owed to its employees. As the result of a case brought by Fidelity Trust & Safe Deposit Company in Philadelphia, the courts placed N&W in receivership in February of 1895. Receivership was a unique tool the courts used in the late 1800s to allow for the reorganization of railroads that were ultimately bankrupt. Because railroads were so important to the public, the courts endeavored to keep them operating even when they were insolvent. The standard procedure was to appoint the existing leadership of the company as "receivers" and create a "new" company with its finances reorganized to satisfy creditors. The receivers appointed for N&W were the company president, Frederick Kimball, and Henry Fink. Fink was a longtime railroad man and had been involved in the receivership of several other railroad companies.

It is not clear whether any of those who filed lawsuits against the company ever received compensation. Many of the lawsuits in the records at the Bedford County Courthouse were marked as dismissed on the same day, May 20, 1895—the same year the company went into receivership. There are no notes to indicate if perhaps a settlement was reached in all the cases or if they were somehow dismissed because of the changes in the company's structure. Norfolk & Western Railroad Company was reorganized and emerged as Norfolk & Western *Railway* in October of 1896. Former president Kimball was appointed as the Chairman of the Board for the new company. Eventually Norfolk & Western Railway joined with the Southern Railway and many other railroad companies over the years to form the Norfolk Southern Corporation.

From Twisted Metal to the "Cannonball"

After the salvage operation for passenger train Number Two was completed at Thaxton, the materials that were left were taken to the company's facilities in Roanoke. Norfolk & Western brought the remnants of engine Number 30 to its Motive Power Department. In the photographs taken of the wreck, the engine was a hardly recognizable jumble of metal. The men at Motive Power went to work. They performed a complete repair and overhaul on the once mighty Baldwin locomotive. N&W records indicate the engine was back in service at least by 1892, and by 1900, it had regained its status as one of the fastest trains in the South. Number 30 joined with a sister engine, Number 35, and shared duty pulling the "Cannonball" from Richmond to Norfolk, Virginia. "Old Number 30" was capable of carrying passengers at speeds reaching seventy-six miles per hour on the Richmond-Norfolk run.[6] The engine that had once smoldered in the waters of Wolf Creek steamed down the rails for twenty-five more years. It was officially retired in December of 1914.[7]

Keeping the Story Alive

As the years passed by and one generation handed the torch to the next, the story of the wreck at Thaxton was revisited locally from time to time. The names of some of the locations would change. Buford became present-day Montvale, and Liberty became Bedford. Locals would occasionally recount what they knew of the story, and the Bedford newspapers would print historical articles periodically over the years. As with most stories, the details can sometimes be lost over time, and as people pass the story on, it can sometimes be embellished or enhanced.

In 1937, nearly fifty years after the wreck, the *Bedford Democrat* printed an article that revisited that terrible night at Thaxton. The details in the article seemed to match up well with historical descriptions of the wreck from 1889. The author used copies of the 1889 Roanoke papers obtained from the sister of James Bruce, the fireman who was killed on the train. One interesting discrepancy in the article is a section describing General Manager Joseph Sands' experience as a passenger on the train. Sands was not on the train but came on a relief train several hours after the wreck had occurred.

The 1937 article was followed nearly fifty years later by a 1981 write-up by Kenneth E. Crouch in the *Bedford Bulletin-Democrat*. Crouch's article touched on the details of the wreck and introduced a new twist to the story, mentioning that there were two wrecks that night. One wreck was that of the passenger train, but the other was a freight train wreck at Little Otter River. Crouch's source for the information was two letters he received from a historian in Roanoke, and they are on file at the Bedford County Museum. Unfortunately, there seems to be no record of any freight wreck at Little Otter on July 2, 1889, and none of the newspapers made any mention of a second wreck that night. In fact, the July 6, 1889 edition of the *Richmond Dispatch* actually refers to damage that was done to the bridge over the Little Otter River by the storm, but the article makes no mention of any wreck at that location.

One of the most interesting twists added to the story appeared in a 1996 *Bedford Bulletin* article written by Rebecca Jackson-Clause. In the article, Jackson-Clause related a story told to her by William A. Otey of Thaxton. William's father, A. B. Otey, was ten years old when the wreck took place and lived on a farm across from the fill at Wolf Creek. The young A. B. Otey and his father were reportedly among the first people to arrive to lend a hand. The story passed down to William Otey by his father was that an old black man named Jabe Showalter had discovered the washout before passenger train Number Two arrived. Showalter heard the train coming and attempted to flag it down at Bocock's Crossing. Otey described the train as "in a hurry" and that it would not stop for Showalter.[8] Based on the story about old Jabe Showalter, is it possible that the tragedy at Thaxton could have been prevented?

Jackson-Clause pointed out in her article that the story of Jabe Showalter "came to light only last week" in her conversation with William Otey.[9] In fact, Showalter's name did not appear in any historical records related to the wreck until the 1996 article. Census, birth, and marriage records do confirm that a Jabe Showalter (sometimes listed as Jake or Jacob) lived in the Thaxton area around the time of the wreck. That particular Showalter was not an old man in 1889 but was likely somewhere in his mid-to-late thirties.

The man who passed the Showalter story down to William Otey was his father, A. B. Otey. His grandfather, A. W. Otey, was most certainly one of the first of the local men to arrive at the wreck to help. A. W. Otey provided a signed statement that was included in Norfolk & Western's report to the railroad commissioner about the severity of the weather the night of the wreck. A. W. Otey was also called as a witness during the 1893 case brought by the family of Will Marshall against N&W. Otey was at the scene when Will Steed's body was pulled from the wreck and was asked about that incident by the attorney representing the Marshall family. Curiously, the Marshall family attorney questioned Otey only about the body of Will Steed. There was no mention of the Jabe Showalter story. The case was tried four years after the wreck, which meant Showalter's attempt to stop the train should have been widely known around town. Marshall's attorneys made several efforts to point out that the train was moving quickly, and it seems that an eyewitness statement from Jabe Showalter would have been included in their case.

It is often difficult to get to the facts in a story that just happened yesterday, much less over 130 years ago. The story of Jabe Showalter did not surface in any of the court testimony, was not mentioned in the newspapers in 1889, and did not surface publicly until about one hundred years after the wreck. These details suggest that the story may have been more folklore than factual. However, there are certainly reasons that the Showalter story could have remained hidden all those years. For example, Showalter could have died or moved away before the court cases were held. If Showalter were not available, it is unlikely that someone else through hearsay could tell his story in court. In addition, the racial climate in the late 1800s would certainly make it possible that Showalter's story was ignored by anyone who might be able to carry it to the proper authorities. Jabe Showalter could very well have been there by the tracks that night, offering one final chance for passenger train Number Two to avoid its fate at Wolf Creek. No one will ever know for sure.

Those who had survived that horrible night at Thaxton continued with their lives. Many of them had lived through the Civil War, and the wreck was one more memory of death and destruction to add to a lifetime of painful experiences. As with any large group of people, their lives would take many different directions through time. They went on to start families, survive later train wrecks, work as missionaries, commit suicide, reside in insane asylums, and become politicians or successful businessmen. The Thaxton train station eventually closed in the 1950s, but the little village is alive and well. Of course, all the passengers and crew eventually found themselves crossing over the river into eternity just as those who had already made the trip in such a terrible fashion that night in 1889. The last known living member of the passengers and crew to leave this earth was Janie Caven, who died sometime after 1965 and most likely in October of 1968. Everyone will make that same trip to the other side someday down the line, and it would be nice to meet the men, women, and children who took their last ride on passenger train Number Two. If this book has served its purpose, you will be able to introduce yourself to any one of them and say, "Hey, I remember you!"

11

SACRED DUST

"We may not know where their ashes slumber, but God himself will keep a faithful watch over their sacred dust."[1]

— REV. C.D. FLAGLER IN A EULOGY FOR THREE
CLEVELAND MEN LOST AT THAXTON, JULY 7, 1889

Chunks of granite lay strewn about the intersection of 8th Street, North Ocoee, and Broad Street in Cleveland, Tennessee. A crumpled vehicle rested against the base of a memorial monument, the driver unharmed but an obelisk which had stood over twelve feet tall was split in two. The entire top shaft, shaped like a miniature Washington Monument, rested on the ground near the base of the marker. The citizens of Cleveland had erected the monument in honor of three prominent citizens who perished in an 1889 Virginia train wreck.

Now, nearly 124 years after the memorial had been placed, another prominent citizen stood at the site and gazed at the damaged monument. His name was Allan Jones (no relation), a successful businessman and lifelong resident of Cleveland. Beyond his personal interest in the history of his hometown, Jones had a special connection to the monument. He had recently purchased America's oldest tailor-made clothing company, Hardwick Clothes, which was founded by a man named Christopher Hardwick in 1880. One of the names inscribed on the shattered monument was John Hardwick, a victim in the wreck at Thaxton and son of Christopher Hardwick.

One month later, in May of 2014, my phone rang. I was at a beach vacation rental with my family. We were staying at the same place I had been three years earlier, when I had decided I would write a book to memorialize the lives lost and long forgotten at Thaxton. The caller was Allan Jones.

Mr. Jones had searched online and found my book, *Lost at Thaxton*. He wanted to learn more about the location of the wreck at Thaxton so that he could visit the site. I explained how he could access the site by pulling off the shoulder of the highway, but there wasn't anything to mark the location. I offered to send him a map link to show the spot. He was surprised to hear that there wasn't any type of marker at the site.

The Virginia Department of Historic Resources sets the criteria and approval process for historical highway markers, but the not-so-trivial cost of manufacturing and placing the marker requires a sponsor to pay for it. I mentioned that I hoped to get a marker at the site, and I had considered starting a crowdfunding campaign to raise the money. I took a sip of lemonade just as Mr. Jones began to reply.

"I'd like to pay for that," he said.

After I wiped the lemonade from my laptop screen, I thanked Mr. Jones for his offer to fund a marker and went straight to work with his team at the Allan Jones Foundation to get the ball rolling. The Virginia Board of Historic Resources only reviewed marker applications at its quarterly meetings and we wanted to get the marker approved and installed as soon as possible. My call with Mr. Jones was on May 27, and the deadline to file the application for the next board meeting was June 1. I knew I was going to be busy with sandcastle construction projects with my kids during the day, so I happily prepared to burn the midnight oil once again.

I drafted text for the marker, completed the VDHR application, and put together some map images for a possible location. The folks at the

Allan Jones Foundation began to reach out to their contacts in Virginia, and after dotting the I's, crossing the T's and crossing the fingers, we submitted the application on May 30, 2014 to Jennifer Loux at the Virginia Department of Historic Resources. She would put the finishing touches on the proposal. Then we would have to play the waiting game until the Board of Historic Resources decided in September.

I received word from Jennifer Loux on September 18, 2014. The historical marker had been approved. Finally, the horrible wreck at Thaxton, one of the worst in Virginia history, would have a permanent marker to commemorate the site and honor the memory of those lost on that terrible night in 1889. It was the most awesome email I had ever received.

While we waited for the Virginia Department of Transportation to survey the site, approve the final location for the marker, and for the Sewah Studios foundry in Pennsylvania to build and ship the marker, we could keep ourselves busy with preparations for a dedication ceremony, scheduled to take place about eight months later.

On the morning of July 3, 1889, one day after the wreck, Christopher Hardwick and Samuel Marshall were desperately scouring the wreckage at Thaxton for the remains of their sons. Marshall had offered $1,000 to anyone who could help him find his son Will's body. After searching for the entire day and finding nothing, the men prepared to take a sorrowful trip home. Hardwick sent a telegram ahead of their departure. "Our last hopes gone. Have been to the wreck and find no trace of the boys."[2]

On May 19, 2015, perhaps for the first time since Hardwick and Marshall had searched for their sons in the debris of passenger train Number Two, a group of citizens from Cleveland, Tennessee arrived at Wolf Creek in Thaxton, Virginia. The group included Allan Jones and his sons, Cleveland Mayor Tom Rowland, Cleveland historians Debbie and Ron Moore, Debbie Riggs, author of a book about the wreck's impact on Cleveland titled, *The Day Cleveland Cried*, and Tommy Hopper, the great-nephew of wreck victim John Hardwick.

This time the visitors from Cleveland came for a moment of celebration, the dedication of the historical marker to commemorate the 1889

Thaxton Train Wreck. No longer would there be "no trace of the boys" at the wreck site as Christopher Hardwick had said in his somber telegram. The marker would serve as a permanent remembrance at the spot where the young men from Cleveland and fifteen others took their final breaths.

We gathered together beneath a tent just a few yards away from the same tranquil creek which had morphed into a raging river and washed away the railroad tracks all those years ago. The Allan Jones Foundation had handled all the arrangements for the ceremony, and along with the folks from Cleveland, several Thaxton residents, railroad history buffs, local media, and a group of Thaxton Elementary School students attended the dedication.

During the ceremony, I had the honor of reading aloud the eighteen names of those who had lost their lives in the wreck at Thaxton. It was something I had hoped to do since the early beginnings of my research for the wreck and a moment I will never forget.

After some words from Mayor Tom Rowland, Allan Jones provided some background on how he got involved in the process. He thanked many of the people on his team who had helped to get the marker in place and coordinate the dedication ceremony. He expressed how proud he was to play a part in getting a marker at the site, and how important the history of the wreck is to Thaxton and the importance of the site to the whole state of Virginia. After he finished, we listened to a beautiful rendition of "Amazing Grace" played on the bagpipes by Major Burt Mitchell.

The ceremony finished with a closing prayer from Randy Martin, pastor of Cleveland's Broad Street United Methodist, the same church where Christopher Hardwick had once worshiped. Just as Pastor Martin finished his prayer, a Norfolk Southern freight train eased by our gathering on the very same route that had once carried passenger train Number Two to that fateful place.

It was like a Hollywood ending to a story that began 126 years earlier with a terrible crash and the horrific deaths of eighteen men, women, and children. Some of their remains were buried together in an unmarked grave in the city of Roanoke, while many of their bodies were never found, consumed by the flames in the aftermath of the wreck. The wreck and those who perished were mostly forgotten, until a confluence of chance circumstances led to that moment. A permanent marker unveiled to memorialize them all as a train passed by on a warm spring day. Reverend Flagler's words from 1889 rang true. God had indeed kept his eye upon that final resting place for their sacred dust.

Thaxton Train Wreck Historical Marker. (Taken by author in May 2015)

12

BIOGRAPHY

"During the past twenty years Mr. Marye was in two serious railroad wrecks, being badly injured in the second. In the first he saved the lives of a number of women and children who were imprisoned in a coach wrecked near Thaxton."[1]

— OBITUARY FOR PASSENGER BURTON MARYE,
RICHMOND TIMES-DISPATCH, DECEMBER 6, 1924

In order to produce as complete a history of the wreck at Thaxton as possible, each of the known seventy-four passengers and crew members aboard passenger train Number Two was individually researched. The goal was to determine exactly who they were, where they were from, where they may have been going, and what happened to them in the years after the wreck. Often in life, the people who are more "prominent" in the public eye have more of their history preserved than those who are not considered as important. Prominence and value do not go hand-in-hand, of course, and each person on the train was researched using the same methods regardless of society's measure of their importance.

Most of the employees working for the Pullman Company on the sleeper cars were listed only by their first initial and last name in

Norfolk & Western's report on the wreck. Without a full name or a hometown, those employees were very difficult to identify with certainty. Only a reference to a Pullman company employee's name or hometown written in a news article in 1889 could help pinpoint exactly who they might have been. The Pullman employee records on file at the Newberry Library in Chicago are composed primarily of records from the 1920s and later. Each Pullman employee on passenger train Number Two was checked in the Pullman Company Archives, but only one was positively identified. That employee was John W. Scott, who stayed with the company until 1929. In the absence of any other information, the origin and destination cities for each Pullman sleeper were cross-referenced with the directories for those cities. In all but one case, those directories contained a match for the Pullman employees on the train, but it is not certain that those were the employees in the wreck at Thaxton.

In some cases, a tremendous amount of information was discovered about a person and their descendants. On the opposite end of the spectrum, even a full name could not be found for some of the passengers and crew. There were still others who could be identified, but the trail would sometimes go cold after the wreck. A few of the single ladies who later married were difficult to track down, especially if there were no records of where they were married or to whom.

There are a few examples in which a person's identity is almost certain, but there was not an absolute confirmation that he or she was the same person who was on the train. If insufficient evidence was found to confirm that a particular person was in fact the one on the train, no background information for them was mentioned in Part I of this book. This chapter passes along any theories or clues related to the identity of passengers and crew. Photographs found of any of the individuals on the train are also included here. Passengers and crew are listed in the order they were riding on the train, from the engine back. The names of those killed appear in *italics*. For additional details and source information related to the passengers and crew, please visit lostatthaxton.com.

Biography 139

Photo of Norfolk & Western Engine Number 37. This engine was one of six ordered on the same date with the same specifications as engine Number 30. (Photograph courtesy of Norfolk and Western Historical Photograph Collection, Norfolk Southern Archives, Norfolk, Virginia. Digital image courtesy of Special Collections, Virginia Tech, Blacksburg, Virginia.)

Photo of Norfolk & Western Engine Number 29. This engine was another of the six ordered on the same date with the same specifications as engine Number 30. (Photograph courtesy of Norfolk and Western Historical Photograph Collection, Norfolk Southern Archives, Norfolk, Virginia. Digital image courtesy of Special Collections, Virginia Tech, Blacksburg, Virginia.)

Photo of Norfolk & Western Engine Number 30. The photograph is not dated but may have been taken in the years following the wreck at Thaxton, after the engine was rebuilt. (Photograph courtesy of Norfolk and Western Historical Photograph Collection, Norfolk Southern Archives, Norfolk, Virginia. Digital image courtesy of Special Collections, Virginia Tech, Blacksburg, Virginia.)

Engine Number 30

Norfolk & Western ordered engine Number 30 in 1886 from the Baldwin Locomotive Works in Pennsylvania. It was originally an N&W class N engine, and later upgraded to a class N1 with larger cylinders and driving wheels. Baldwin's classification for the engine was an 8-30C, and the common description for this type of engine was a 4-4-0 "American" engine. The 4-4-0 designation was from the Whyte Notation System used for classifying steam engines. It simply meant that there were four wheels grouped together on the front, followed by four larger driving wheels and with no wheels behind those driving wheels.

The "American" designation was chosen because so many engines of this type were used in America.

The following souls were aboard engine Number 30:

Photo of James Edgar Bruce's grave marker. Note the inscription reference to his death at Thaxton. (Taken by author in April 2012 at Chestnut Hill Baptist Church Cemetery in Big Island, Virginia)

Biography

James Edgar Bruce
Resided: Roanoke, Virginia
Born: November 16, 1866
Died: July 2, 1889

James was the fireman on the train, in charge of feeding coal to the engine. Based on reports at the time, Bruce's body was never found. He had just renewed his $1,000 accident policy the week before the wreck, and his family may have used some of that money to purchase the memorial stone pictured above. He was just twenty-two years old and the son of J. C. Bruce. The Bruce family had recently moved to Roanoke from Lynchburg at the time of the wreck. His brother, Sam Bruce, became a freight conductor for Norfolk & Western.

Photo of Patrick Donovan's grave marker. Note the inscription references his death on N.&W.R.R. near Thaxton. (Taken by author in September, 2012, at Holy Cross Cemetery, Lynchburg, Virginia)

Patrick Donovan
Resided: Lynchburg, Virginia
Born: March 1, 1857
Died: July 2, 1889

PATRICK WAS the engineer and was the only one of the three on the engine who had positive identification of his remains. Donovan's silver watch was used to identify his body. His family sued N&W for damages, but the results of the case were not in the files at the Bedford County Courthouse.

Census records, cemetery records, and Lynchburg City Directories record that his father was likely Patrick Donovan Sr., a grocer in Lynchburg. Engineer Donovan was a member of St. Patrick's Beneficial Society and was described in the July 3 edition of the *Daily Virginian* as "a popular and highly esteemed young Irishman of this city, a wholesouled active man, always at his post."[2] Donovan was thirty-two years old. His remains were buried in what is known today as the Holy Cross Cemetery in Lynchburg.

Photo of Patrick Donovan's grave marker. (Taken by author in September, 2012, at Holy Cross Cemetery, Lynchburg, Virginia)

Alvin M. James
Resided: Petersburg, Virginia
Born: About 1857
Died: July 2, 1889

Alvin was the Road Foreman of Engines in the Lynchburg Division of Norfolk & Western. He was traveling on a personal trip the night of the wreck, but he had volunteered to help on the engine due to the treacherous weather conditions. At the time of his death, Alvin left behind a wife, Annie M. James, a four-year-old daughter, Sallie M. James, and a nearly two-year-old son, Clyde B. James.

Tragedy would strike the James family again very soon after the wreck. For reasons unknown, Annie James passed away just a month after the wreck took her husband, and little Sallie and Clyde never knew their parents. Their grandparents, Annie's father, William H. Wheary, and his wife, Sallie, raised them.

Postal Car Number 280

United States Railway Post Office (RPO) car Number 280 was carrying sixty packages of letters, fifteen sacks of papers, and three registered pouches. The following souls were aboard:

James J. Rose
Resided: Abingdon, Virginia
Born: About June 1870
Died: July 2, 1889

At the time of his death, James was assisting postal clerk Lewis Summers and had an eye on making a career for himself. He had long been pursuing his true love, Lillian May Grubbs, and had promised her he would depart from his wild ways if she would take his hand in marriage. She had agreed, and they were to be married the day after the wreck. After his death at Thaxton, there is no clear information to determine whether Lillian ever married anyone else or not. James's brother, L. Wood Rose, was living in Knoxville at the time of the wreck.

Lewis Preston Summers
Resided: Abingdon, Virginia
Born: About 1868
Died: December 10, 1943

Lewis Summers had been on the job as a railway postal clerk only a little over a month when he was caught up in the wreck at Thaxton. Unlike his coworker, Summers survived the wreck and managed to escape with cuts and bruises but no major injuries. Within a year after the wreck, Summers was appointed as the postmaster in Abingdon. Over the next several years, he became a lawyer and authored several historical books about the southwest Virginia area.

He eventually became the United States District Attorney for the Western District of Virginia. It was in this role that he became involved in a scandal that resulted in a short prison term for him. In 1924, a federal grand jury indicted Summers on six counts for making false claims that totaled $398.75. The claims were recorded for "personal service" rendered by a Miss Hattie Perkins, a former clerk in his office.[3] He stood accused of signing off on three payments of $166.25 for services that were not actually rendered for the office.

Testimony during the trial revealed that the married Summers had set up an apartment for Miss Perkins in Washington, DC. The landlady who managed the apartment testified that Summers had visited on more than one occasion and had given her an alias of Henry Wilson. He also led her to believe that Hattie Perkins was his daughter. Lewis fought the charges tooth and nail, including appeals to the Supreme Court and a request to President Calvin Coolidge for a pardon. The conviction stood, however, and he was sentenced to fifteen months in an Atlanta penitentiary in 1927. He served nine months before being paroled. Summers continued his legal career and was considered one of the most successful trial lawyers in Virginia at the time of his death. Lewis also continued to work as a historian and was appointed a member of Virginia's State Historical Commission.

Lewis Preston Summers. Reproduced from Lewis Preston Summers, History of Southwest Virginia, 1746–1786, Washington County, 1777–1870. (Richmond, VA: J. L. Hill Printing Company, 1903)

ETV&G Baggage and Express Car Number 57

The baggage car that belonged to the East Tennessee, Virginia, and Georgia Railway carried suitcases, personal effects of the passengers, an express safe with valuables amounting to several thousand dollars, and even baskets of chickens. Just about everything in the car was destroyed, including the contents of the safe. The human cargo aboard the baggage car was as follows:

Robert H. Ashmore
Resided: Mossy Creek, Tennessee (Now known as Jefferson City)

Robert was the express messenger for the Southern Express Company and escorted the packages entrusted to the shipping company. He was seriously injured in the wreck and at first was rumored to have been killed. After doctors treated him at Granville Sanitarium in Liberty, he appears to have returned to his job as an express messenger. The 1891 Chattanooga City Directory lists a Robert H. Ashmore, messenger for the Southern Express Company.

Records seem to indicate that the wreck at Thaxton was not Ashmore's only brush with death on the Norfolk & Western. On Christmas Eve, 1897, an R. H. Ashmore was the express messenger on passenger train Number Six when it collided with a freight train near Pulaski, Virginia. A postal clerk on the train was killed instantly, and, as at Thaxton, it was thought that Ashmore might die due to head and chest injuries. It appears that Ashmore survived and continued in his job as an express messenger. The 1908 Chattanooga City Directory also listed a Robert H. Ashmore as a messenger for the Southern Express Company.

In 1923 Ashmore's life may have ended tragically. A death certificate for Robert H. Ashmore, born at Mossy Creek, Tennessee, indicates that he took his own life. The certificate lists his occupation as an express freight agent. At fifty-seven years of age, Ashmore killed himself with a .38 pistol shot to the chest.

There is no solid link between the Robert H. Ashmore who was in the wreck at Thaxton and the other records related to the later wreck and eventual suicide. The circumstantial link is the common name, occupation, and locations.

James Calder Cassell
Resided: Roanoke, Virginia
Born: March 16, 1856
Died: September 29, 1936

James Cassell's presence on passenger train Number Two was crucial because he documented many of the events that led up to the wreck. Cassell compiled much of the report that Norfolk & Western put together concerning the accident. He continued to work for N&W until 1905, when he resigned his position as assistant to the president for health reasons. After leaving the company, he focused on entrepreneurial efforts, including a ration and commissary contracting business called Cassell & Elliott and the Roanoke Overall Company. Cassell was raised in Mount Joy, Pennsylvania, and married Emma Boyer in 1883.

James Calder Cassell. Reproduced from George S. Jack and Edward Boyle Jacobs, History of Roanoke County (Roanoke, VA: Stone, 1912)

Captain William Henry Ford
Resided: Lowry, Virginia
Born: About 1839

Baggage Master Ford sustained significant injuries in the wreck. He and his wife's fight with N&W management and their eventual lawsuit are documented in Chapter 8 in this book.

Second-class coach Number 54

The second-class coach was designated the "smoker" for gentlemen who wished to have a smoke while they were on the train. It was also the car for passengers like Robert Davis who were not allowed a place elsewhere on the train because of their race. The following souls were documented aboard the coach at the moment of the wreck:

Robert Davis
Resided: Eastville, Virginia
Born: About 1845

Robert was born in Smithville, Virginia, and had been a part of the African Methodist Episcopal Church since 1869. He became an elder for the church in 1874, and his role of presiding elder kept him traveling to churches in the AME Danville District. His work brought him aboard passenger train Number Two the night of the wreck. He was one of the early leaders of Bethel AME Church in Eastville, which still exists today.

W. C. Glass
Resided: Roanoke, Virginia, or Lynchburg, Virginia

Glass was one of two brakemen and was stationed at the front of the train. Norfolk & Western's report listed Glass from Roanoke, but the *Daily News* in Lynchburg listed Glass as a Lynchburg resident. He suffered a broken arm and was scalded about his head and face in the wreck. There is a 1907 obituary for a William C. Glass of Roanoke. The obituary mentions that this particular William C. Glass died in Roanoke but was a native of Lynchburg, where his remains were taken. It also mentions that at the time of his death he was a yard conductor for Norfolk & Western and that he had been with N&W for twenty years. All of that information makes it possible that this was the same W. C. Glass who was part of the wreck at Thaxton, but there is no definite link.

Joseph Goldberg
Resided: New York, New York

Goldberg suffered bruises to his leg and shoulder and had an injured hand. Other than Goldberg's hometown, there was little information published about him.

William Graw
Resided: Knoxville, Tennessee

William was an extra express messenger on the train working with Robert Ashmore. His plan was to learn the route and someday take a job with the Southern Express Company. He apparently jumped from the coach as the wreck occurred and suffered a sprained ankle and scalding to his hands and face. There is some confusion related to Graw's name. The N&W report listed him as W. H. Graid. His hometown paper, the *Knoxville Journal*, listed him as William Graf in a July 3, 1889, article, but subsequent articles in the same paper referred to him as William Graw. A July 26 article mentions that Graw was seen around town with badly burned hands and lists his address as 13 East Depot Street. The city directory for Knoxville in 1889 confirms that a William Graw lived at that address.

Roland P. Johnson
Resided: Roanoke, Virginia
Born: About 1856
Died: May 1908

Although scarring on his hands would remain as a reminder of that night, train conductor Roland Johnson recovered from his injuries sustained in the wreck. He continued to work as a conductor for Norfolk & Western until 1900, when the company promoted him to the position of trainmaster for the Scioto Valley division at Portsmouth, Ohio. The trainmaster position required him to manage the movement of trains and personnel for his division. The job was challenging and stressful, but a bit safer than riding the rails.

Roland died at Hot Springs, Arkansas, eight years after taking his new job. He had a kidney disease, which at the time was generally called Bright's Disease. Many people traveled to Hot Springs hoping that the warm and radioactive waters would provide relief for a number of ailments. The springs were not able to save him. He left behind his wife, formerly Lelia Allison from Glade Spring, Virginia, and a son who was also working for Norfolk & Western in Portsmouth. Roland's father was L. F. Johnson, from Bristol, Tennessee.

W.C. Meyers
Resided: Roanoke, Virginia

Meyers was listed in the N&W report as a newsboy from Roanoke. The report recorded his injury as a gash across the forehead, but no other information was provided to help identify him. The 1888 and 1889 Roanoke city directories contain only one Meyers, an Israel Meyers, listed as a lumberman. There are several men with the last name spelled Myers, but none matches the newsboy's initials. He may have been a young man living in the household of one of these other Meyers, but there is no way to be sure.

Frank DeWitt Tanner
Resided: Lynchburg, Virginia
Born: About 1860

Tanner decided to have a late night smoke, and for him that was the only difference between surviving the wreck at Thaxton or succumbing to the death and flames that ended the life of his friend John Kirkpatrick. Tanner was at home in Lynchburg within a couple of days after the wreck. He was badly bruised and needed a cane to support his sprained ankle. Norfolk & Western's report listed Tanner with two different names. In a table listing the injured persons, he is listed as F. V. Tanner. However, Superintendent James Cassell referred to him as F. D.

Tanner in his official statement in the same report. Cassell praised Tanner's help during the rescue effort.

Records suggest that Cassell had the correct initials for Tanner. The 1887 Lynchburg City Directory lists a Frank D. Tanner, part of the firm Tanner, Bliss & Co. The company was a grinding operation for the mineral baryte. The Bliss in the firm was actually Tanner's cousin, George H. Bliss.

One mystery that remains unsolved is exactly why Frank Tanner was with John Kirkpatrick on his trip to Roanoke. As detailed in Part I of this book, Kirkpatrick had made the trip to resolve an issue with the banks that had resulted from a check endorsement for a friend. The Lynchburg papers revealed the purpose of Kirkpatrick's trip but did not release the name of the friend who had written the bad check. Is it possible Tanner was the one who had caused Kirkpatrick to take the fatal trip, or was he just conducting business of his own in Roanoke that day?

Within a few years after the wreck, Frank moved to Washington, DC, along with his mother, Sarah Jane. His cousin, George Bliss, also relocated to the nation's capital with his mother and family. Frank was managing a small hotel there called The Norman, possibly named after his father, Norman S. Tanner. As of 1900, Frank was still single, and by 1910 he no longer appears in census records. His cousin and former business partner, George Bliss, died in 1903.

First-class Coach Number 63. (Photograph Courtesy of Delaware Public Archives)

Interior Layout of Coach Number 63. (Courtesy of Delaware Public Archives)

Randy L. Goss, Photo Archivist at the Delaware Public Archives, could not verify the interior photo is of Coach 63, but the order number printed on the photograph is the same as that stamped on the exterior photo of Coach 63. Ten cars were purchased in that order. Mr. Goss confirmed that "If this photo is not #63, #63 would have looked like this."[4]

First-class coach Number 63

ACCORDING to Norfolk & Western's report to the railroad commissioner, at least twelve of the eighteen known to be killed were located on coach Number 63. This report does not include John Dowell, whose location on the train was unknown and was never mentioned in the N&W report. The *Knoxville Journal,* on July 4, 1889, suggested that the three men from Cleveland, Tennessee, were not on this coach, but were in the smoker instead, second-class coach Number 54. The official report from the railroad listed the men on the first-class coach. Additional evidence that the three men were on the first-class coach came from conductor Roland Johnson. He was on the second-class coach and reported seven passengers aboard at the time of the wreck. That count matches the railroad's report, assuming Johnson was including himself in that number.

Death was indiscriminate in the first-class coach, taking men, women, and children. It was believed that most of them were not killed in the initial crash but were crushed to death when the Pullman sleeper Beverly slammed down on top of it a few minutes after the wreck. It is possible that some were still alive and trapped when the fire broke out, but in general the railroad leadership took the position that it was unlikely. Virginia Railroad Commissioner James Hill stated, "I am satisfied, though, that no person was burned up alive. They were all dead before any fire reached them."[5] There were others, however, who referred to groans heard from the wreckage even as the fire broke out. Rescuers worked around the Beverly sleeper in particular to free those pinned down there, but underneath that rubble lay coach Number 63. If anyone had still been alive and trapped when the fire broke out an hour and a half after the wreck, they were probably on this coach.

Nathan Cohen
Resided: Roanoke, Virginia
Born: August 22, 1863
Died: July 2, 1889

Nathan immigrated to the United States from Bremen, Germany, in August of 1881 aboard the S.S. *Weser*. After arriving in America, he spent his first six years living in Baltimore near his uncle before moving to Roanoke. Cohen's Uncle Aaron Hess and his family were Nathan's only relatives in the United States. Census records from 1900 reveal that those relatives were Aaron, his wife Dinah, and their sons Louis and Nathan. The July 5, 1889, edition of the *Baltimore Sun* reported that Cohen was on the smoking car, but the official report from N&W listed him on the first-class coach. His friends described him as "cordial, genial and kind."[6]

Fred T. Dexter
Resided: Beverly, Massachusetts
Born: About 1866

Fred was one of the "Fortunate Four," the only four men who survived the wreck on coach Number 63. Norfolk & Western listed sixteen people on the coach, and twelve of them were killed. This meant that the coach had a 75 percent death rate among those known to be on board. Like fellow travelers Alpheus Wilson and Lewis Summers, Dexter felt that the train was going at a high rate of speed. Like those two men, he was also asleep when the train wrecked.

Dexter gave his estimate of the train's speed in an 1890 deposition for his lawsuit against N&W. He was seeking $10,000 compensation for his injuries in the wreck, which were listed as an injured spine, back, and internal organs. The Norfolk & Western report a year earlier had documented his injuries only as a strained left shoulder and various cuts and bruises. In his testimony, Dexter estimated the train was going about thirty miles per hour, but he clarified that the estimate was based on the force of the impact.

Because one of the wheels of the railroad car had ground into his back and side during the wreck, Fred felt the force of the impact severely. He was confined to his room for six weeks while he recovered from his injuries. He did not return to his business as a traveling shoe salesman until three months after the accident. Extra-long breaths would cause him severe pain, and occasionally just while talking, he would double over with pain in his side. He also complained of having a nervous condition. A doctor named S. N. Jordan (likely Seth Jordan from census records) in Columbus, Georgia, examined Dexter in 1890 and provided a perfect example of late nineteenth-century language. He described Dexter as "being naturally a splendidly formed man having an unusually fine chest."[7] Dr. Jordan felt that Dexter had some lung damage from the wreck, and he did not think that Dexter "will ever be the man again that he was before the accident."[8]

Records indicate that Fred was probably from Nova Scotia and born on November 7, 1865. The 1890 city directory for his hometown of Beverly listed Dexter at 21 Mulberry and mentions he was a salesman in Newburyport. Dexter's deposition recorded that he worked for Burley & Usher in Newburyport. It is not an absolute fact that the Dexter at 21 Mulberry is the same as the one in the wreck, but it seems likely.

That same Fred Dexter from 21 Mulberry appears on an 1894 naturalization record that lists his birth date and home country. Marriage records in Massachusetts record that Frederick Torrey Dexter eventually married Harriet Shaw Brown in 1894 before moving to Minnesota. There he would start a family and continue his career in the shoe business until his death on March 8, 1919.

Robert Brent Goodfellow
Resided: Roanoke, Virginia
Born: About 1867
Died: September 3, 1921

Like Fred Dexter, Robert Goodfellow was one of only four men to survive the crash in coach Number 63. He too was a young man in his early twenties. At the time of the wreck, Norfolk & Western employed him as a clerk in Roanoke. Goodfellow was on his way home to visit his family in Washington, DC. He somehow managed to escape through the roof of the coach and suffered a sprained ankle along with a number of cuts and bruises.

Robert was the son of the late Major Henry Goodfellow, a former Union Army soldier and Judge Advocate in the US Army. After spending time as a clerk for N&W in Roanoke, Robert returned to his home in Washington, DC, to work as a civil engineer. Robert had attended Notre Dame and was also considered one of the top tennis players on the local circuit in the nation's capital.

In the summer of 1896, things began to go horribly wrong for Robert. Neighbors heard screams and the sound of a gunshot outside the home of his mother in the early evening of June 26. Crowds gathered outside the home, and the police arrived to investigate. His mother would not let the officers go upstairs in her home, and she did not wish for the officers to make any arrests. The rumor was that Robert had come home and was recklessly brandishing his revolver.

It was not long before violence again called on the Goodfellow household. Four months after the first incident, Robert savagely attacked his brother, John. Even though at times Robert would seem entirely normal, it appeared that he was dealing with some type of insanity. He was having episodes in which he would become obsessed with mathematical equations and mysterious signs. The assault on his brother resulted in a diagnosis of insanity, and Robert spent the last twenty years of his life at St. Elizabeth's Hospital for the Insane. Medical records from St. Elizabeth's note that Robert had been involved in a railroad accident and had "received a wound on the head," but the records do not indicate whether doctors thought the injury contributed to his medical diagnosis.[9]

John Millard Hardwick
Resided: Cleveland, Tennessee
Born: August 14, 1856
Died: July 2, 1889

Hardwick and his two friends from Cleveland were well-known and well-liked young men in their hometown. John had founded Cleveland Stove Works with his father, Christopher, and his brother, Joseph. The company was already prospering when John was killed, and it continued to be a very successful business, later named the Hardwick Stove Company. In 1981, Maytag acquired the Hardwick Stove Company.

In her book *The Day Cleveland Cried: A History of the Monuments*, Debbie Riggs described a heated controversy involving John's brother and the monument that the citizens of Cleveland had erected for the three men killed at Thaxton. In 1911, the United Daughters of the Confederacy were given permission to place a new Confederate memorial in downtown Cleveland. The location selected for the new memorial was just feet away from the monument for Hardwick, Marshall, and Steed. It had been over twenty years since the wreck, and a suggestion was made to relocate the monument for the three men to the Fort Hill Cemetery in town. Steed and Hardwick's families agreed with the idea to move the monument because it had at that point "served its purpose."[10]

The family of the third Cleveland man killed in the wreck did not agree. Will Marshall's mother stated, "I didn't get his body, and it made me sick at heart to think about its removal. It's the only monument to his memory."[11] John Hardwick's brother, Joseph, ignored her protest and hired several men to disassemble the monument. The Marshall family obtained an injunction to block removal of the monument, but workers took it down anyway. It rested on the side of the street in Cleveland for several months.

While the legal battle went on, the citizens of Cleveland took up sides. Those who supported the move based their support primarily on the size of the new Confederate memorial. It was much larger than the existing monument for the men killed at Thaxton and, as a result, made the smaller tower seem insignificant next to it. While the memorial to the Thaxton deaths remained disassembled on the street, the new Confederate monument was raised and dedicated at a town celebration.

John Hardwick's sister, Nora, was emotionally distraught over the treatment of her brother's marker. She decided that she would pay to have it restored to its original position next to the new Confederate tower. As the men she hired worked to reconstruct the monument, her brother, Joseph, cursed at them and ordered them to stop or he would have them arrested. Nora decided she would sue her brother, and her case was combined with the Marshall family's case against him. In testimony, Joseph initially argued that he was the only caretaker for the monument and that he was concerned that it would not be maintained after his death. He also cited the size of the new Confederate memorial as another reason to have his brother's marker moved. Under cross-examination, Joseph revealed the likely reason for his desire to move the monument. His house was directly across from the monument, and he explained, "I never wanted it there, for I did not want my brother's grave yard almost in front of my front door."[12]

The Marshalls and John's sister, Nora, won the case, and both monuments stand side-by-side to this day in downtown Cleveland.

John Kirkpatrick
Resided: Lynchburg, Virginia
Born: About 1865
Died: July 2, 1889

The tragic details of Kirkpatrick's unplanned trip to Roanoke that resulted in his death at Thaxton are detailed in Part I of this book. He was the son of Reverend James M. Kirkpatrick and the nephew of Major Thomas J. Kirkpatrick, a prominent attorney in Lynchburg. He was also the secretary of the Young Men's Democratic Club. The club placed a special tribute of respect for Kirkpatrick in the July 7 edition of the *Daily News* in Lynchburg. In addition to praising John for his intelligence, the tribute described John by saying, "That so conspicuous were his sincerity of purpose and his generosity of heart that he will ever be held in affectionate remembrance by all who knew him."[13]

James W. Lifsey
Resided: Roanoke, Virginia
Born: Possibly around 1860
Died: July 2, 1889

Although he was an employee of Norfolk & Western, James Lifsey was described in the N&W report to the railroad commissioner as "on a pleasure trip."[14] In that report, Lifsey resided at Roanoke, but it does not appear that he had been living there for long. He is not listed in either the 1889 or the 1888 Roanoke city directory. An article in the July 4 edition of the *Daily Virginian* stated he had been living in Petersburg, Virginia, but had transferred to the Lynchburg division recently. In addition, a few different sources report that his family and original hometown were in Greensville County, Virginia.

Lifsey's name was one of the more inconsistently reported names in the newspapers. His last name was generally printed as either Lifsey or Livesay, but in at least one case it was listed as Lipsey. Newspapers also inconsistently reported his first name as James, John, or even Joseph. The official report from N&W listed him simply as "J. W. Lifsey."[15] The edge in the evidence seems to be that his name was James Lifsey. Census records from Lifsey's home county of Greensville for 1860, 1870, and 1880 consistently show a James Lifsey, son of John H. These same census records also note the spelling as Lifsey in the Greensville County area.

None of the newspaper articles written around the time of the wreck reported that Lifsey was married or that he had any children.

Dennis Mallon
Resided: Roanoke, Virginia
Died: July 2, 1889

Early reports of the wreck mentioned that Mallon was on his way to New York to be married, but there was no additional information printed concerning his pending marriage. Details of exactly who Dennis was are sparse. The 1889 Roanoke city directory listed him as the janitor at the N&W general office building, which was also his residence. He was not included in the 1888 directory and apparently moved to Roanoke from elsewhere.

William Franklin Marshall
Resided: Cleveland, Tennessee
Born: 1867
Died: July 2, 1889

Most of Will Marshall's story is told elsewhere in this book. Marshall's family continued to fight for him many years after his death. In 1893, the family became the first plaintiff to win a case against N&W and was awarded $7,500 by a jury. The decision was overturned on appeal a year later. His family's battle with John Hardwick's relatives over the removal of the monument for the three Cleveland men is detailed under the entry for John Millard Hardwick in this chapter. Marshall was the son of Samuel W. and Mary Marshall.

Charlene Peyton
Resided: Radford, Virginia
Born: Possibly December 6, 1888
Died: July 2, 1889

Charlene was the youngest victim in the disaster at Thaxton and perished in the flames along with her mother and father. She was listed simply as "Chas. Peyton, (child)" in Norfolk & Western's official report.[16] Charlene was never identified by name in the newspaper accounts of the wreck but was instead typically listed as the child or infant of Mr. and Mrs. Charles Peyton. She is listed by name in *The Peytons of Virginia II Volume Two*.[17]

There is also an entry listing a "Charline Peyton" in Virginia birth records on December 6, 1888.[18] The father on that record is listed as Chas. L. Peyton, and the mother as Josie E. Her mother's name was Jessie, so while it is not certain, it is possible that this entry is for Charlene. The record indicates she was born at Alexandria, Virginia, which was an area that her parents had previously lived.

A small heart found in the wreckage was generally believed to be Charlene's and was buried with the other remains at the city cemetery in Roanoke.

Charles Llewellyn Peyton
Resided: Radford, Virginia
Born: May 4, 1865
Died: July 2, 1889

Charles was born in Orange County, Virginia, and was the son of Thomas Jefferson Peyton and Sarah Elizabeth Reynolds. As mentioned in Chapter 3, Charles was working as a stenographer for Norfolk & Western. He had seven siblings, which included his brother George, who also worked in the railroad business with the Virginia Midland Railway.

Jessie Hacquard Peyton
Resided: Radford, Virginia
Born: About 1863
Died: July 2, 1889

Jessie was the second daughter of Louis Hacquard of Wheelersburg, Ohio, part of Scioto County. Her friends and family had to endure a pair of heart-wrenching telegrams as they waited for confirmation of exactly what had happened to Jessie and her family. An initial telegram received in Wheelersburg brought news that they had been killed and no bodies were found. A second telegram arrived and brought a small glimmer of hope that they would at least be able to give a proper burial to Jessie and Charlene. It reported that their bodies had been recovered. Eventually they would get the terrible confirmation that the original message was correct.

It does appear that Jessie's family may have put up a memorial marker for her. A photograph taken at the Wheelersburg Cemetery in

Ohio and posted on the website waymarking.com shows a marker matching her information, including a death date of July 2, 1889. The marker is part of a headstone that includes her father, Louis. The inscription on the marker reads "Rest, Darling Sister, Rest. We Will Meet You Again."[19]

William C. Steed
Resided: Cleveland, Tennessee
Born: July 1857
Died: July 2, 1889

Steed's body was the only one among the casualties that was recovered from the wreckage untouched by the flames. His father was J. C. Steed, a brick contractor in Cleveland. Several newspapers reported that Steed was carrying $14,000 with him, but the *Roanoke Daily Times* reported a sum of $600. There is no additional evidence to confirm the accuracy of those reports or any explanation as to why he was carrying that large a sum with him. Will was unmarried and left no wife or children behind.

William C. Steed. (Courtesy of the History Branch and Archives of Cleveland Bradley County Library)

John I. Stevenson
Resided: Richmond, Virginia
Died: July 2, 1889

Details of John's business trip with his boss John Bowers and the search for his whereabouts can be found in Part I. John had been married twice. His first wife's maiden name was Shanks, and they had two sons together. She passed away, and John married his wife at the time of his death, Ada Smith of Henrico County. City directory and census records identify John's two sons as John I. Stevenson Jr. and William A. Stevenson.

John Sr. was also a member of the same Presbyterian church in Richmond where fellow passenger Pattie Carrington's memorial service was held.

J. Fred Temple
Resided: Chicago, Illinois
Born: About 1865
Died: July 21, 1895

J. Fred Temple. (Courtesy of Nono Burling, Great-Great Niece of J. Fred Temple and the Burling Family)

Temple was another of the "Fortunate Four" men who survived in coach Number 63. He was described as "a tall, fine looking young man, with dark-brown moustache and dark eyes."[20] His back was injured and his head "severely cut," and he needed the assistance of friends to help him walk when he returned to Chattanooga on July 4.[21]

Despite being spared in the wreck at Thaxton, Temple still suffered an untimely death just six years later. He drowned while working as a civil engineer in South McAlester, then part of Indian Territory and now part of the state of Oklahoma. There were at least two conflicting reports about how he drowned. One report described a fall from a bridge into the water below. A second report suggested that he drowned while bathing in the reservoir. He left a wife and two children at the time of his death.

Harry B. Wheeler
Resided: Richmond, Virginia
Born: November 3, 1857
Died: July 2, 1889

Many details of Harry's life are covered in Part I, including information about his wife Laura and five-year-old son Arthur Becker Wheeler. Likely because Harry knew the risks of the excessive amount of railroad travel required for his job, he carried quite a bit of life insurance. He had a $5,000 accident policy purchased by his employer, the Union News Company. Wheeler also had a second $5,000 life insurance policy. The wreck had taken him from his family, but he left behind some resources for them to carry on.

Laura used some of those resources to return to their hometown of Baltimore after Harry's death. They had named their son, Arthur, after Harry's boss, Augustus Becker. Arthur eventually became a dentist. Census records over the years showed that Laura and Arthur remained close. In 1920, Laura was living with Arthur, his wife, and son. By 1940, Arthur was divorced and living at the same address as his mother. It does not appear that his mother ever remarried.

J. A. Young
Resided: Radford, Virginia

Norfolk & Western reported that Young suffered a sprained ankle and bruises to his body. Beyond that, very little is known of the fourth man that survived the wreck in coach Number 63. The N&W report listed him as a train dispatcher out of Radford. No matching city directory records or census records for Young help to identify him. The official report from N&W listed him as A. A. Young, but newspapers at the time of the wreck identified him differently. One article listed him as J. B. Young, and other articles referred to him as J. A. Young. There is a J. A. Young listed as a train dispatcher many years later in the 1898 Roanoke city directory, but nothing in particular links him to the wreck at Thaxton.

"Beverly," Pullman Sleeping Car Number 289

Norfolk & Western documented the spelling for the name of this Pullman sleeper as Beverly, but in the book, *A Century of Pullman Cars*, historian Ralph Barger lists the spelling as Beverley. The car number in Barger's book, 289, matches the one recorded by N&W. Railroad officials also referred to Beverly as the Norfolk sleeper because of its intended destination. The circumstances of the wreck gave the Beverly sleeper a central role in the way the night unfolded at Thaxton. Immediately after the train went into the washout, Beverly was still partially supported by the embankment and did not fall to the bottom. Minutes later, the earth below Beverly gave way, and the heavy sleeper crashed down on the first-class coach, which likely caused many of the deaths.

W. J. Barksdale
Resided: Richmond, Virginia

Barksdale provided one of the more memorable quotes from the accident when he declared to the other ladies in Beverly, "I will not leave, ladies, until I see you all safe; I am a Virginian."[22] He was listed from Richmond in Norfolk & Western's report and in at least one newspaper account. Both of those sources referred to him as W. J. Barksdale, but an article in the *Dallas Morning News* listed him as W. I. Barksdale. Richmond census and city directory records do not provide any clues to his possible identity. He emerged from the wreckage uninjured and acted heroically to assist the ladies of Beverly out of the demolished sleeper.

T.B. Bott
Resided: Richmond, Virginia

Bott was one of the men in Beverly who worked with Barksdale to help free the ladies trapped inside the sleeper. He was most likely Thomas B. Bott from Richmond, but that still could have been one of two men. A senior and a junior lived in Richmond at the time, and both were at an age that made it possible for them to be on the train. The July 6 edition of the *Daily Virginian* listed the name as T. B. Scott, and there was a Thomas B. Scott living in Richmond in 1889. The official report from N&W listed him as T. B. Bott and described him as uninjured in the wreck.

Janie Caven
Resided: Dallas, Texas
Born: About 1871
Died: After 1965

Several of the ladies on the train that night did all they could to render aid and ease the suffering of those around them. Janie's efforts were especially memorable and received special mention in numerous newspapers and witness accounts. The *Salem Times-Register* described her efforts after the accident:

> "She labored untiringly in their rescue and in caring for them when rescued, tearing the clothing from her person into strips as bandages for their wounds. By such self-forgetfulness, such fortitude in peril and blessed ministrations, she has won an endearing place in the hearts of all who can appreciate a true and noble womanhood."[23]

Fellow passenger, Fred Temple, witnessed Janie's work and felt that her name should be "written in history upon a page of gold."[24] Temple also summarized what many felt about Janie after the wreck. "The world don't hold many nobler, better women than that."[25]

A record in the Social Security Death Index suggests that Janie was probably born on September 11, 1871 and died in Washington, DC, in October of 1968. It does not appear that she ever married, and at least as late as 1965, she was living in Washington, DC, with her sister. When she died, she was the last known person who experienced that terrible night in 1889 to take that memory with her.

Edmund L. DuBarry
Resided: Crewe, Virginia
Born: March 17, 1842
Died: December 4, 1908

N&W Superintendent Edmund DuBarry was one of the last people pulled from the wreckage before the flames erupted, consuming the train and all those who remained inside. DuBarry and fellow Superintendent James Cassell's presence on the train enabled many details of the events leading up to the wreck to be preserved. He continued to work for Norfolk & Western as a superintendent until his death in Norfolk, Virginia, in 1908.

His friend and former coworker, James Cassell, was one of many N&W employees who attended funeral services for DuBarry. An obituary eulogized that "He was a man of large ability and stood high in the estimation of the management of the Norfolk and Western Railroad."[26] DuBarry took one final train ride to Baltimore for burial after the funeral services at Norfolk. He left a widow, Laura, and his daughter, Louise.

W. H. Haywood
Resided: Norfolk, Virginia

Haywood worked for the Pullman Company as the porter on the Beverly Sleeper. Little information was recorded for him, and even the correct spelling for his name is in question. Norfolk & Western listed him only as "W. H. Haywood, colored porter on sleeper Beverly," and he is included with the list of the uninjured.[27] Two separate newspaper articles refer to him as W. H. Hayward and mention a shoulder injury and that he was badly bruised. There are no records in the Norfolk city directories around that time that match with any of the spellings of his name. Pullman Company records also provided no clues to his identity.

Burton Marye
Resided: Richmond, Virginia
Born: About 1862
Died: December 5, 1924

Marye was one of several men who worked together to free the ladies trapped inside the Beverly sleeper. He injured his wrist slightly when he forced his way out of a window. Norfolk & Western listed Marye as uninjured. When he was involved in another railroad accident twenty years later, he was not quite as fortunate.

On December 15, 1909, Marye was a passenger on Southern Railway Train Number 11. Around 6:30 a.m., the train was crossing a trestle over Reedy Fork Creek just north of Greensboro, North Carolina, when it hit a broken rail. Three coaches and two sleepers were hurled down over twenty feet into the stream below. Twelve people were killed. Marye's leg was broken, but he once again had survived a terrible accident.

It seems as if Marye may have been just a little bit accident-prone from an early age. A short statement printed in the *Alexandria Gazette* when Marye was about thirteen years old foreshadowed his future—more newsworthy accidents. "Burton Marye, a little son of Col. Morton Marye, fell from a tree yesterday afternoon and was so unfortunate as to have his arm broken. It was only a month ago that he broke his other arm."[28]

Walter C. Masi
Resided: Norfolk, Virginia
Born: About 1856
Died: October 9, 1914

Walter was in the drugstore business in Norfolk with his brother, Fred Masi. His brother had died of tuberculosis at Salem, Virginia, just outside of Roanoke, only one day before the wreck. There is no information available to determine whether Walter had gone to Salem to retrieve his brother's body or if he was returning to Norfolk from some other location. Walter suffered a sprained ankle in the accident and continued to work in the pharmaceutical industry until his death in 1914.

Roberta B. Powell
Resided: Marshall, Texas
Born: December 27, 1848
Died: November 10, 1899

Roberta Powell had been visiting Salem, Virginia, near Roanoke with her daughter, Inez Sparkman, and Inez's friend, Janie Caven. Her injuries from the wreck were listed as "General stiffness and soreness of limbs and body" by Norfolk & Western.[29] Roberta was one of several passengers who filed a lawsuit, and in her case, she was seeking $4,000 in damages. The case was marked as dismissed in 1895 along with most of the others.

John Thomas Rowntree
Resided: Knoxville, Tennessee
Born: About 1854
Died: March 13, 1930

Hardware buyer John Rowntree was back on his way to New York only twelve hours after the wreck. Norfolk & Western reported Rowntree as uninjured in its report to the railroad commissioner. Things apparently got worse for Rowntree once he arrived in New York. Upon his arrival, he found that he could barely walk and was bedridden for days. Back issues and "nervous prostration" caused him to return to Knoxville and miss a month of work due to his injuries.[30] The pain did not subside for several months. He ended up quitting his job in Knoxville that October and relocated to Denver.

Rowntree filed a lawsuit against Norfolk & Western seeking $4,000. It was listed as dismissed on May 20, 1895, along with most of the other cases on record at the Bedford County Courthouse. Eventually he would start his own hardware company, John T. Rowntree, Inc. with headquarters in Los Angeles and branch offices in San Francisco, Seattle, Salt Lake City, Denver, and Mexico City. Rowntree was originally from Spartanburg, South Carolina. He died in 1930, leaving behind his wife, two daughters, and a son.

John Thomas Rowntree. Reproduced from John Steven McGroarty, Los Angeles from the Mountains to the Sea, Volume 2. (Chicago and New York: The American Historical Society, 1921)

Inez Sparkman
Resided: Marshall, Texas
Born: March 19, 1871
Died: March 16, 1904

Inez was traveling with her mother, Roberta Powell, and her friend, Janie Caven. The young eighteen-year-old suffered the worst of the three, with injuries listed as "Concussion of spine; retention of urine; and disturbance of menstrual and digestive functions."[31] Like her mother, Inez sued Norfolk & Western and requested $10,000 in compensation for her injuries. The case was marked dismissed the same year as many of the others, 1895.

Inez returned to her hometown in Texas and married a doctor from South Carolina, Hugh R. Carwile. There she gave birth to a son, Hugh G. Carwile in 1898. Unfortunately, Inez died just before her thirty-third birthday in 1904.

John W. Scott
Resided: Portsmouth, Virginia
Born: December 28, 1858
Died: July 1, 1946

The man in charge of the Beverly Sleeper was also one of the most active in the rescue efforts once he was freed. His efforts received special recognition from N&W Superintendent James Cassell in his report to General Manager Joseph Sands. The official report listed Scott among the uninjured, but news reported in his hometown mentioned that he was badly bruised when he returned to Norfolk.

Pullman Company employee records reveal that Scott remained a conductor with the company until his retirement in 1929. A headstone located in Olive Branch Cemetery in Scott's hometown of Portsmouth, Virginia, lists Scott's middle name as Wonycott and his death on July 1, 1946. There is no direct evidence that the headstone is for the same John W. Scott who was the sleeper conductor that night, but there are several very strong links. His wife Maggie's name also appears on the headstone, and the cemetery contains several others that match up with family names associated with Scott in the census records over the years. Finally, the date of birth listed for John Wonycott Scott on the headstone is an exact match with the birth date on file in the Pullman Company employee records.

"Toboco," Pullman Sleeping Car Number 416

Norfolk & Western recorded the name for this sleeper as Toboco, but historian Ralph Barger's book, *A Century of Pullman Cars*, suggests Toboca may have been the name for the car instead. Toboco was described as a through car originating out of Chattanooga and headed to Washington, DC. As was common, the sleeper was often referred to by its destination and in this case was called the Washington sleeper by some.

Pattie Carrington
Resided: Houston, Texas
Born: September 12, 1879
Died: July 2, 1889

Many of the heartbreaking details of Pattie's life and her father's painful string of tragedies are mentioned in Part I of this book. When Pattie died in the wreck, her father Allen had lost two children and two wives within a span of about ten years. His misery continued even after Pattie's death. One year after the wreck at Thaxton, Allen's third wife gave birth to another baby girl, Elizabeth. Elizabeth would die before reaching the age of one. Pattie's father met with his own tragic death just three years after she was killed at Thaxton. While visiting family in Signal Mountain, Tennessee, Allen was riding in a horse-drawn buggy across a bridge constructed with logs. The horse slipped, and Allen jumped from the buggy, hoping to save himself. Unfortunately, he

Pattie Carrington. (Courtesy of Jim Hutcheson)

landed on his back and was paralyzed completely. He died within a few hours of the accident.

W. H. Craig

Craig was listed as a "colored porter" working on the Pullman sleeper car, Toboco.[32] Norfolk & Western provided no hometown or other information for Craig. Toboco was headed for Washington, DC, and the 1890 city directory did list a William H. Craig as a porter. No other evidence confirms whether this was the same man as the porter on Toboco.

Annie Fishpaugh
Resided: Knoxville, Tennessee
Born: About 1862

Annie survived the wreck uninjured and continued to work as a milliner in Knoxville for several years afterward. The Knoxville city directory listed Annie as late as 1895, but then the trail grows cold. She does not appear in later directories in Knoxville, nor is she listed in later directories in her hometown of Baltimore. It is not clear whether something happened to Annie after 1895 or if she married and subsequently changed her name.

Her mother, Rebecca, had married John Sheckells when Annie and her sister Emma Fishpaugh were younger. In Rebecca's 1918 obituary, there is no specific mention of family member names to point toward what might have happened with Annie.

L. H. Garnett

Like fellow porter W. H. Craig, N&W listed Garnett as a "colored porter" working on the Toboco sleeper.[33] The 1889 Washington, DC, directory lists a porter named Lewis Garnett, but similar to Craig and the other porters, there was not enough information recorded to estab-

Edith Hardester
Resided: Knoxville, Tennessee
Born: August 1862

Like her friend, Annie Fishpaugh, Edith was a single lady from Baltimore and worked in Knoxville as a milliner in 1889. Norfolk & Western listed Edith as uninjured in the wreck. Within a year or so after the wreck, Edith married Jordan Tucker, and they had two sons. It appears that Edith and Jordan may have moved around a bit while Jordan worked as an insurance agent. A death notice in the January 30 edition of the *Sun* suggests that she may have died in her hometown of Baltimore in 1917. No clear evidence links the Edith Tucker mentioned in the notice with Edith Hardester.

John Irby Hurt
Resided: Abingdon, Virginia
Born: December 1, 1866
Died: March 28, 1935

Generally referred to as J. Irby, he and his sister Rosa Lee escaped from their wrecked sleeper unharmed. He went on to become an attorney and marry at least twice. He married his first wife, Margaret Fulkerson, in 1891, but she died in 1894. At his death in 1935, his widow and one daughter survived him, most likely Marcella Daoust.

Rosa Lee Hurt
Resided: Abingdon, Virginia
Born: December 23, 1868

Rosa Lee was one of the ladies who received special mention in the newspapers of 1889 for her help rendering aid to the wounded at Thaxton. Three years after the wreck she married James Bell, who would become the president of the First National Bank of Abingdon.

During a telephone conversation in March of 2012, Rosa Lee's granddaughter, Anne Hutton of Abingdon, provided a personal look at her grandmother. Anne said that her grandmother did not like to speak of the wreck, but one of her aunts had told her how she had torn her petticoat to help wrap up the injured. Rosa Lee was nearly naked by the time the relief train arrived. She ultimately became a "wonderful grandmother" to Anne, and she described her as an accomplished artist.

Rosa Lee also had quite the sense of humor as well. When Anne was younger, her parents would let her take the car just about anywhere as long as her grandmother was along for the ride. At what Anne described as "the most inappropriate times," Rosa Lee would take her cane and press the accelerator to the floor while her granddaughter tried to keep the car under control!

Irene Baley Jackson
Resided: Knoxville, Tennessee
Born: December 14, 1863
Died: January 18, 1955

A Knoxville newspaper advertisement for her employer, Lon Mitchell, described Irene as an "Artistic Lady from Baltimore."[34] Her hometown was Cambridge, Maryland, on the peninsula of Maryland's Eastern Shore. On the night of the wreck, she was on her way to be married in her hometown. Fortunately, she was uninjured, and she became Mrs. Eugene H. Wilkes just two weeks later on July 17, 1889.

The couple settled in Eugene's hometown of Laurens, South Carolina, and they remained there the rest of their lives. They had at least three sons and four daughters. When Irene died in 1955, the only known survivor of the wreck still alive was Janie Caven.

Rida Jourolmon
Resided: Knoxville, Tennessee
Born: March 8, 1863
Died: November 22, 1930

In 1889, Rida was a schoolteacher in Knoxville. After 1900, Rida became a missionary for the Southern Presbyterian Mission in China. She never married and continued her work as a missionary until 1927. She died three years later in Montreat, North Carolina.

H. T. Moss
Resided: Philadelphia, Pennsylvania

Philadelphia city directories for 1889 and 1890 do not provide a conclusive link to who the H. T. Moss was on the train. There are two possibilities in the 1889 directory, a Henby T. Moss and a Henry Moss. The 1890 directory includes Henby T. Moss again, along with a Herbert Moss. The lack of a middle initial for Henry and Herbert means they cannot be ruled out as possibilities.

Catherine Lightfoot Carrington Thompson (Aunt Kate)
Resided: Dallas, Texas
Born: May 26, 1825
Died: June 25, 1893

As you read the many accounts of Aunt Kate's sorrow that night at Thaxton and in the weeks afterward, the full effect of the devastation and pain caused by the wreck can still be felt. Her tormented cries for Pattie carved their way into the memories of many of those who shared those terrible moments with her. One paper that listed the injuries sustained by some of the passengers even included "deranged with grief" in its description of her injuries.[35] An article in the *Roanoke Daily Herald* reported that her grief had "well nigh deprived her of reason."[36]

The loss of her great-niece, Pattie Carrington, devastated Catherine, who had always loved Pattie just as if she had been her mother. In a letter written to his sister when Pattie was still a baby, her father spoke of Aunt Kate's devotion and mentioned that she described Pattie as "the child of her old age."[37]

The official report from Norfolk & Western mentioned that Catherine was bruised around the eyes and that she had stiffness and soreness around her eyes and neck. The bruising must have been very significant. Fellow passenger Edith Hardester actually thought that Catherine had lost an eye, but that was not the case. One of the family members who brought Catherine back from Roanoke after the wreck was Joseph Chappell Hutcheson. In a letter to his daughter written just over a week after the wreck, Hutcheson described Aunt Kate as "terribly bruised" and that her eyes were "black as my lad from a blow on the temple, but no serious or dangerous injury."[38]

Catherine Lightfoot Carrington Thompson. (Courtesy of Jim Hutcheson)

Her greatest injury was to her heart. Hutcheson wrote that when they found Aunt Kate, she was no longer able to shed tears, but she could only emit "a dry, pitiless wailing grief."[39] Catherine died nearly four years later and was buried in Richmond's Hollywood Cemetery. Her final resting place is directly beside Pattie's grave.

Alpheus Waters Wilson
Resided: Baltimore, Maryland
Born: February 5, 1834
Died: November 21, 1916

Alpheus Waters Wilson. Reproduced from Carlton Danner Harris, Alpheus W. Wilson A Prince in Israel (Louisville, KY: Board of Church Extension of the Methodist Episcopal Church, South, 1917)

It took some time for Bishop Wilson to heal after his harrowing night at Thaxton, but he was back to his travels within a month of the wreck. About a week after the accident, his daughter, Nina, described him as "covered with bruises—is a pale yellow tint."[40] The terrors of that night stayed with him as well. She described his mental state a month later in a letter to Collins Denny, the man who met Alpheus in Roanoke after the wreck. "There is no use in denying his nerves are much shaken. I believe it will be months before he is himself again."[41]

Wilson would overcome his nerves and place himself in peril many more times on the rails and at sea over the remaining years of his life. Only one year after the wreck, he set out on a trip around the world to visit the missions set up by the Methodist church overseas. During the twenty-seven years he lived after he escaped the waters of Wolf Creek, he made his way to China, Japan, London, Toronto, and numerous locations throughout the United States. He and his wife narrowly avoided other train wrecks in the years after Thaxton. In one incident, Alpheus and his wife were on their way from San Francisco to a conference in New Orleans. A bubonic plague outbreak had taken place in San Francisco. Authorities turned Bishop Wilson and his wife away at the Texas border because they did not have a certificate of health from a doctor. This setback proved fortunate for the Wilsons because the train they were removed from went on to wreck before reaching New Orleans.

Alpheus may have been the most widely known of any of the passengers or crew on N&W train Number Two. His travels took him around the country and around the world. When he died in 1916, news of his death was printed in newspapers from New York to California. Carlton Danner Harris, a biographer for Bishop Wilson, wrote, "He is destined to be recognized as a great historical character that has been a tremendous factor in shaping the history of Methodism on this continent and in the regions beyond."[42]

"Calmar" Pullman Sleeping Car

Calmar was on its way to Washington, DC, with some passengers who had boarded the sleeper in New Orleans. As mentioned in Part I, Calmar was touted as part of the "Montezuma Special" that connected the capital cities of Mexico City and Washington, DC.[43] Calmar was the only car on the train not to fall into the washout, but the raging fire made short work of the luxurious sleeper. Fortunately, since the fire did not start right away, there were no injuries to passengers or crew on Calmar.

A. Banks

Banks was one of two porters on Calmar. Like all of the porters, N&W provided only first initial and no hometown information for him. An Albert Banks was listed as a porter in the Washington, DC, city directory for 1890.

Sophie Boutron
Resided: New Orleans, Louisiana

The world of high fashion was not limited to the young milliners on the Toboco sleeping car. Sophie oversaw the Dressmaking Department at the D. H. Holmes department store in New Orleans. Mrs. Boutron had a bit more experience than the young ladies on Toboco. She appeared as early as 1861 in the millinery business in New Orleans.

Samuel Lee Boyd
Resided: Lynchburg, Virginia
Born: About 1863

Just before Wolf Creek swallowed train Number Two, Boyd was relaxing on a stool at the rear of the Calmar sleeper. He sprang into action almost immediately when disaster struck and was one of the most active men in the work to get help and assist with the rescue efforts.

In 1889, Boyd was still living with his father, who owned a candy store and "ice cream saloon" in Lynchburg.[44] Some evidence suggests Boyd later married and moved to Newport News, where he died in 1933 and left behind his wife, Mabel, and son, Harry.

William H. Cooley
Resided: New Orleans, Louisiana

The N&W wreck report recorded Cooley as W. H. Cooley from New Orleans. A separate passenger also mentioned that Cooley was from New Orleans in his account of the wreck. The only match in New Orleans city directories in 1889 and 1890 was William H. Cooley. He was a clerk for Shattuck & Hoffman, a cotton broker in the Crescent City.

Cooley no longer appeared in New Orleans directories after 1891. Just after the wreck, the *Washington Post* had reported that Cooley had relatives in Washington, DC, but there are no records to show that Cooley relocated there either.

P. P. Dounsberry
Resided: San Antonio, Texas

Dounsberry's name seems to be a possible error in Norfolk & Western's list of passengers. There are no dounsberry, downsberry, or dunsberry listings in San Antonio directories in the years around the wreck. There was a Phineas P. Lounsberry in San Antonio at the time, and one of the Roanoke papers did list him as P. Lonsburg. Phineas was the owner of the St. Leonard Hotel on the Main Plaza in San Antonio. N&W did misspell some of the names in its report, but it is only speculation that Lounsberry may have been the passenger on Calmar that night.

J. H. Elam

No other person was more active in the rescue efforts than J. H. Elam. Just before the train collapsed into the fill, Elam was walking toward Toboco. If he had made his move just a few minutes earlier, he may well have been thrown into Wolf Creek, and he would have needed rescue himself. Norfolk & Western provided little information on Elam despite all of his efforts to help free passengers and his eventual six-mile run to Buford to get help. N&W identified him only as a furloughed baggage master. Elam's actual identity remains a mystery.

An 1891 article in the *Roanoke Times* mentions a Joe Elam who worked as a passenger conductor for N&W between Roanoke and Lynchburg, but there was already a conductor named Joe Elam living in Lynchburg prior to the wreck. It seems unlikely that he would have been a baggage master after already holding the conductor position.

J. P. Gage
Resided: Fairfield, Iowa

N&W listed Gage's hometown as Fairfield, Iowa. He may have been on his way to Washington, DC, to file a patent application for a folding chair concept he devised. James P. Gage of Fairfield, Iowa filed an application on July 10, 1889. It was an improvement to an 1887 design also patented by Gage. The new design allowed for a rocker style chair and included a padded upholstered seat. Gage's patent was approved and issued later that year, on December 24, 1889.

It is not absolutely certain that the James P. Gage who filed the patent was the passenger on Calmar, but there is strong evidence. An 1885 Iowa state census in Fairfield lists only one possible J. P. Gage, and the family information recorded in that census is a perfect match with an 1880 record for Gage. That same year Gage received approval for a patent on a harrow device he had created—an implement used in farming and similar in function to a plow. The application listed Gage's full name as James Pike Gage. He would have been about forty-three years old when the wreck occurred at Thaxton.

The problem with stating with certainty that the J. P. Gage on Calmar was the same man filing the patents is the length of time between the 1885 Iowa census and the wreck. It is certainly possible that another J. P. Gage had moved to Fairfield during that four-year period, but the circumstances suggest that the inventor James Pike Gage was the man on the train.

E. Gambler

Along with A. Banks, Gambler was one of two porters on Calmar. N&W did not list a hometown, but if Gambler worked out of Washington, DC, where Calmar was headed, he may have been Edward Gambler. The DC directory listed Edward as a porter in 1890.

Henry N. Martin
Resided: Chattanooga, Tennessee

Norfolk & Western recorded Henry Martin's hometown as New Orleans, Louisiana but articles and an interview published in the *Chattanooga Daily Times* indicate that Martin was a resident of Chattanooga, working there as an insurance agent. The 1889 Chattanooga city directory also lists Martin as an insurance agent there. The discrepancy in hometowns reported may have been because Martin had only recently moved to Chattanooga.

According to a mention in the New Orleans paper, *The Weekly Pelican*, Henry N. Martin lived in New Orleans and moved to Chattanooga around 1887 to engage in business there. That man was Henry Neill Martin, a Confederate Civil War veteran and one-time candidate for congress in New Orleans.

Martin made it through the wreck at Thaxton unscathed, but just one year later he may have died from a kidney disease that was then known as Bright's Disease. An obituary printed in the July 11, 1890, *Chattanooga Daily Times* mentions that the popular local insurance agent had passed away the day prior, leaving behind his wife and four children.

George A. Masters
Resided: Philadelphia, Pennsylvania
Born: December 5, 1869

George was fresh out of college in 1889. He had graduated from Swarthmore College with an engineering degree in June. After the wreck, he lived in Philadelphia with his father, David, for a few years before starting a wholesale shoe business in 1898. Two years later, he married a woman named Anna Todd.

Masters was unharmed in the calamity at Thaxton, but he experienced his own tremendous loss later in life. In 1912, his son, George Jr., was riding his bike to school on an October Tuesday morning. As he peddled across the intersection of Lincoln Drive and Springfield Avenue in the Chestnut Hill area of Philadelphia, George Jr. was hit by a car. The car was driven by a chauffeur who was taking the son of a local banker to school as well. Witnesses felt that the chauffeur was driving carefully, but that information offered no comfort to George Masters, who had lost his eleven-year-old son.

Charles Montague
Resided: Bristol, Tennessee

Whatever was ailing Charles Montague had been bothering him months before the wreck. A brief statement in the May 9 edition of the *Knoxville Journal* mentioned that Charles was taking a furlough from his job as a Pullman conductor. Two months later, he was still on sick leave the night he traveled on Calmar. A night in the rain probably did not help, but there is no conclusive information about what happened to Charles in the years afterward.

Pauline Payne
Resided: Knoxville, Tennessee
Born: June 10, 1870
Died: August 24, 1900

As a nineteen-year-old, Pauline's efforts to care for the wounded were so notable that she received specific praise in several articles that detailed the events of the wreck. In contrast, the dark side of human nature was on display when someone stole her satchel containing money and jewelry while she ministered to the injured. Her brother, Reuben, was on the train with her, and he continued toward New York after the wreck and then on to a three-month tour of Europe. There was no mention of whether Pauline continued on to her destination as well.

In 1896, she married Edward Maynard, an insurance agent in Knoxville. They had two daughters, Elizabeth and little Pauline. Edward and his two little girls lost their mother to scarlet fever on August 24, 1900. Pauline was only thirty years old at her death and passed away less than two months after her brother had died.

Reuben S. Payne Jr.
Resided: Knoxville, Tennessee
Born: March 1, 1872
Died: June 30, 1900

It might seem strange that a seventeen-year-old was setting off on a three-month tour of Europe, but Reuben and his sister, Pauline, had a number of resources at their disposal. Their father was a former mayor and prominent businessman in Knoxville. In the years just after the wreck, Reuben first worked as a clerk in a clothing store owned by his father. He eventually started a dry goods and furnishing store in Knoxville with several partners. They operated under the name Brown, Payne, Deavers & Co.

Reuben and his sister apparently remained close over the years. He married Mary Towns Gaines just a few months after his big sister exchanged her vows in the summer of 1896. When Pauline returned via ship to New York after her honeymoon, Reuben and his new bride were there at the docks to meet them. In June of 1900, Reuben and his wife were living with Pauline and her family in Knoxville when the census taker came to visit. Reuben died from tuberculosis at the end of that month at just twenty-eight years of age. His sister joined him in death two months later.

Miss Van Keen

The most mysterious listing among the passengers and crew in Norfolk & Western's official report was that of Miss Van Keen. The report listed her as an uninjured passenger on Calmar, but no hometown or other information was included. The name Van Keen does not appear to be a common one. It may be that it was either a misspelling or possibly a duplicate entry or relative of the similarly named Florence Vanuxem.

Florence Vanuxem
Resided: Philadelphia, Pennsylvania
Born: August 25, 1868
Died: January 30, 1949

Florence was traveling on the train with her friends, Pauline and Reuben Payne. Her connection to Pauline in particular strengthened in the years after the wreck. Florence's father had been a successful businessman in Philadelphia, and her brother Louis continued in those footsteps. He became a millionaire and when he died in 1903, he left a substantial inheritance to his sisters, including Florence.

A little over a year after her brother's passing, Florence married a man from Knoxville named Edward Maynard. If that name sounds familiar, that may be because you read the biographical entry for her friend and fellow passenger, Pauline Payne. Five years after Pauline's death, Florence married Pauline's husband and became a mother to her little girls.

H. L. Williams
Resided: Johnson City, Tennessee

The earliest available directory for Johnson City was published for the year 1908, nearly a decade after the wreck. Because of the 1921 fire that destroyed most of the 1890 census records, there is a large gap of time before and after the wreck that makes it difficult to determine H. L. Williams's identity with certainty. The 1908 Johnson City directory had only one possible candidate, Harry L. Williams. Harry was a special examiner for the pension office of the US government, but there is no specific evidence that he was the H. L. Williams in Johnson City in 1889.

Location Unknown

The exact location on the train was not known for at least three passengers and one crew member. The crew member was a Pullman sleeper conductor who oversaw two sleepers, and there was no specific record of which car he was on at the moment the train wrecked. Norfolk & Western listed two of the passengers as location unknown and the third was John Dowell, a man who was not listed at all in the N&W report.

J. T. Castleman

Castleman was the man in charge of the last two sleepers on passenger train Number Two, Toboco and Calmar. The Norfolk & Western procedure for documenting Pullman employees was to list only the initials and last name with no hometown. This practice, of course, created the same issue with the other Pullman employees that made it difficult to determine exactly who they were. Toboco and Calmar were headed for Washington, DC, and there was one listing in the 1889 city directory for a conductor named James T. Castleman. There is no additional information to determine whether that James Castleman was the man on the train at Thaxton.

John T. Dowell
Resided: Rome, Georgia
Born: About 1828
Died: July 2, 1889

An entire chapter in this book covers the story of John Dowell and his undocumented presence on the train. Dowell was a farmer, originally from Virginia, and had moved to Rome with his wife, Lavinia, several years prior to the wreck. When the wreck occurred, his wife had already passed away, and Dowell himself was around seventy years old.

He may have been aboard the sleeper called "Toboco." Edith Hardester, a passenger on Toboco, told the *Sun* newspaper in Baltimore that "the gentleman in front of me was killed" in the same interview that she described young Pattie Carrington also being killed.[45] Pattie's death and location on Toboco was recorded by Norfolk & Western, but no other death in any of the sleeper cars was ever listed. Edith may have been mistaken, or John Dowell or some other unlisted person was the man she saw meet death that night. Toboco had many passengers that had come through from the ETV&G railroad, which was the road that Dowell would have taken out of Rome.

John H. Hager
Resided: Estillville, Virginia

The name of Hager's hometown was changed from Estillville to its current name, Gate City, about a year after the wreck. The 1880 census for Estillville listed a John J. Hager, but there is no listing for a John H. or any information to suggest that John J. was the man traveling on the train.

Judge Henry Solon Kane Morison
Born: June 12, 1846
Died: November 9, 1899

Judge Henry Solon Kane Morison. Reproduced from the Virginia Law Register, ed. George Brian, vol. 9, May 1903 to April 1904 (Lynchburg, VA: J. P. Bell Company, 1904)

Norfolk & Western's report did not list Judge Morison's hometown, but it was most likely Estillville, Virginia. Morison was born and died in Estillville and appeared there in the 1880 census. He was a circuit court judge for the state of Virginia in 1889, and in 1894, he was the Democratic nominee for Congress from Virginia's Ninth District.

He was married to Annis Kyle, and they had six children together. In November of 1899, the judge suffered a stroke while he sat with his family in his library. Morison was a devout Christian and was known to quote scripture in court. In one particular case, the attorney on the losing end said in jest, "If Judge Morison would read more law, and less scripture, he would make a good judge."[46]

ACKNOWLEDGMENTS

In the year 2022, the technology needed to construct a time machine does not yet exist. In the absence of that technology, many different people are needed to preserve and provide access to the records that document our history. The story of the wreck of passenger train Number Two at Thaxton could not have been told without the efforts of those people. Corporations, public institutions, and private individuals all have had a hand in preserving and providing access to the documentation that makes a book like this possible.

Many different sources of information were used during the research for *Lost at Thaxton*. Digital access to newspapers, books, census records, city directories, and death and marriage records was extremely helpful. I am grateful we live in a time when services such as the Library of Congress Chronicling America project, ProQuest, GenealogyBank, FamilySearch, Newspapers.com, and Google Books are available for research.

Electronic tools are still not so mature that they are the only means necessary to dig deep into the details of history. Several collections preserved at libraries around the country were valuable sources of information. Thank you to the Virginia Historical Society for its review of the John Irby Hurt papers for any information related to the wreck. I would like to thank Anne Benham and Margaret Downs Hrabe of the University of Virginia for their help locating microfilm for my research and for providing information from the Collins Denny and Lewis Preston Summers papers. Thank you to Sherrie Bowser of the Virginia Tech University Libraries for her help in reviewing the Norfolk & Western Railway collection. Jo Ellen McKillop Dickie of the Newberry

Library in Chicago and Christine Windheuser with the Smithsonian were both amazingly helpful in my search for Pullman employee and sleeper car records. Sara Logue helped with a search of the Alpheus W. Wilson papers archived at Emory University, and Rebecca Russell of the Woodson Research Center at Rice University was very helpful during the research of Pattie Carrington, Allen Carrington, and Catherine Thompson (Aunt Kate).

The Norfolk & Western Historical Society has a singular focus on preserving the history of the Norfolk & Western Railway. I learned much from the daily messages shared on the society's mailing lists, and I am grateful to Ron Davis and Jim Blackstock for providing some background information for the book. Jim went the extra mile to answer my questions, and I appreciated his time and his help.

I also appreciate the service provided by Tony Teal and the Tennessee State Library and Archives. Their microfilm copy service allowed me to comb through the Tennessee newspapers at my own pace and seeded my own personal microfilm collection.

No city did more to honor its own victims of the wreck than Cleveland, Tennessee, did. That same civic spirit was evident when I worked with Debbie Riggs and Treasure Swanson. Debbie's book, *The Day Cleveland Cried*, chronicled the wreck's impact in Cleveland, and she graciously searched through her own research to answer questions for me. Treasure Swanson manages the History Branch and Archives of the Cleveland Bradley County Library and personally reviewed records at the History Branch, providing fantastic pictures of the monument and Will Steed.

Several others were instrumental in the location and acquisition of photographs that helped to memorialize the wreck at Thaxton. Many thanks go to Jennifer McDaid of Norfolk Southern Corporation for providing several great photos of the wreck, Hotel Roanoke, and engine-related photos. I am also very appreciative for the help of Marc Brodsky, Kira Dietz, and the Virginia Tech Special Collections team for providing the online ImageBase project and for their assistance in getting digital copies of the Norfolk Southern images for the book. Dyron Knick, Archivist at the Virginia Room in the Roanoke City

Library, was also a great help in obtaining one of the wreck photos. Additional thanks go to Randy Goss of the Delaware Public Archives for the photos and background information for coach Number 63 and for the archives' efforts to make their photo collections available online.

Sometimes individuals may have no idea how something they publish online might be useful to others, and that was likely the case with Jim Hutcheson and Nono Burling. Jim posted materials on the "Find a Grave" website that helped connect several dots related to the lives of passengers Pattie Carrington and Catherine Thompson. Nono's public album of old family photos helped me find a photograph of one of the passengers and provided an opportunity for some very enjoyable correspondence about her family history. Both Jim and Nono were gracious enough to respond to my "out of the blue" requests, and they provided photos that helped to put a face with a name for passengers on train Number Two. I am grateful to them both for sharing a little of their family history with the rest of us.

Harold Davey of the Peyton Society of Virginia provided me with background information for the Peyton family that was killed on the train, and most importantly he provided baby Charlene's name so that it could be properly preserved in these pages.

Special thanks go to George Beths of Holy Cross Catholic Church in Lynchburg, Virginia, who gave me "better than GPS quality" directions to the family plot of the train's engineer, Patrick Donovan.

I would like to thank several individuals in Washington County, Virginia, for their help in the research of the four Abingdon citizens who were involved in the wreck at Thaxton. Thank you to Dona Lee, clerk at the Washington County Virginia Circuit Court, for her help in the search for lawsuit records related to the wreck. I also appreciated the help of Melissa Watson and Jane Oakes in the search for photographs of the Abingdon passengers. One of the highlights of this project was my phone conversation with Anne Hutton of Abingdon, the granddaughter of one of the passengers. Ms. Hutton was the only person I spoke to who had actually talked with someone who was on the train that night, and that was a special treat for me.

"Some guy is here researching an old train wreck" was a phrase I frequently heard during my work on the book. Thank you to Linda Steele at the History Museum of Western Virginia for allowing me to peruse the papers of William Creighton Campbell, the man who conducted the service over the mass grave of the wreck victims in Roanoke. Thank you as well to the staff at the Jones Memorial Library for assisting me and providing access to the microfilm for the Lynchburg papers of 1889. Cris Wilson, Reference Librarian at the Kershaw County Library, was no doubt tired of my seemingly endless requests for interlibrary microfilm loans from other states. I very much appreciated her assistance and time, and I am grateful to the Camden Archives for use of its microfilm machines.

The "train wreck guy" was also welcomed at Bedford Museum and Genealogical Library on multiple occasions. David Waite and Manager Doug Cooper were both very helpful, and David located the bell clapper from engine Number 30 in storage at the museum.

The Bedford County Circuit Court was a treasure trove of primary source information, and I want to thank deputy clerks Karen Glover and Meg Kiely for their help both remotely and on-site. Karen was especially gracious and did some preliminary research to confirm for me that they had the records I was looking for ahead of time.

A number of people in Bedford County were very helpful in the effort to put this book together. John Barnhart's excellent article in the *Bedford Bulletin* in 2012 helped many people learn about the wreck in their own backyard. The article also helped spread the word about the ongoing research for the book. Many people reached out with information, and I especially want to thank county residents Dana Gillenwater, James Weeks McCabe, Jim Morton III, and Cindy Neely.

I would also like to express my gratitude to Shirley Dooley and Gene and Peggy Jones for their donations toward the research and publishing efforts.

The early draft readers for *Lost at Thaxton* were Gene Jones, Peggy Jones, Angela Mayfield, Robert Mayfield, Joyce McDonald, and Todd McDonald. They gave their time and energy to help take the book from very "rough" drafts to a final product that contained far fewer missing commas, dangling participles, verb tense issues, confusing passages, and other crimes against the English language. It is not an attempt at flattery to say that any remaining issues with the text are a result of my own subsequent changes or stubbornness to change something that they likely pointed out to me. Todd was a very valuable sounding board throughout the writing process. He also mastered the art of feigning interest in the latest obscure nineteenth-century trivia item I would frequently pass on during an otherwise enjoyable lunch.

The publishing services team at CreateSpace did a fantastic job of taking the manuscript and assembling a professional book design around it. I could not have asked for a better editor than Joseph. He scoured the text so thoroughly that I began to wonder if he might have been a passenger on the train that night in 1889. His attention to detail helped to put the final shine on the words in this book.

In the years after the first publication of *Lost at Thaxton*, many people reached out via email with feedback about the book, questions concerning the wreck, or stories about their own family connections to some of those on board. I love to hear from people about the book and I want to extend my thanks to Tom Heinrich, who contacted me to let me know about Ralph Barger's book and the alternate names for two of the Pullman sleeper cars that were involved in the wreck at Thaxton.

Another one of those who contacted me after the release of *Lost at Thaxton*, was Allan Jones of Cleveland, Tennessee. Allan's call led to a truly monumental achievement. Less than a year after that call, we dedicated a historical marker at the site of the wreck to memorialize all those lives that had been lost at Wolf Creek in Thaxton. I am exceedingly grateful to Allan Jones and the Allan Jones Foundation for everything they did to see that we got a proper memorial to remember the men, women, and children who were killed in the wreck at Thaxton.

I am so thankful to the most incredible wife a man could ever imagine possible. Supportive is not a sufficient word to describe her dedication to this book. She has lost untold hours with her husband behind computer screens and microfilm machines. Fortunately, she did not keep track of the number of times she has read and reread chapter after chapter after the tiniest of changes. She cheerfully read aloud and listened to passages from the book so many times that she may never reclaim the space from her memory banks. She has endured photographs of steam engines, wreck scenes, and passenger coaches taped to the walls at my desk for a time much longer than any woman should ever be forced to tolerate. While most husbands are engaged in some more traditional pastime like golf, hunting, or fishing, her husband decided he would write a book about a 120-year-old train wreck. Most women would question that man's sanity, and rightfully so, but she has believed in me always. I thank her for her sacrifices and her unconditional love.

Finally, in any effort I undertake, I would be making a terrible mistake if I did not share the role of Jesus Christ in my life. Jesus's gift to me is not some supernatural blessing on my work. He walked the earth as a man to teach and show how God intended for us to live. He died on the cross to give us the assurance that we can be right with God, no matter what the most just consequences should be for the choices we make in our own free will. Jesus's life and death have enabled me to trust Him when it comes to living my life in this world and to have confidence that He has paid for my admission into the next.

NOTES

1. Rain

1. *Report of the Railroad Commissioner*, 1890, p. xxxix.
2. "Meteorological Report for June," *Richmond (VA) Dispatch*, July 2, 1889.
3. "Some Texas Weather," *Dallas Morning News*, July 2, 1889.
4. Ibid.
5. "Through Johnstown," *Philadelphia Inquirer*, June 14, 1889.
6. Jefferson, *Notes on the State of Virginia*, p. 29.
7. Peaks of Otter Chapter, Daughters of the American Revolution, *Bedford Villages Lost and Found*, p. 23.
8. Huntington and Latimer, *The Road-Master's Assistant*, p. 67.
9. *Report of the Railroad Commissioner*, 1890, p. xxxix.
10. Ibid., p. xxix.
11. Ibid., p. xxxiv.
12. Ibid., p. xxxii.
13. Ibid., p. xxxiv.

2. Departure

1. *Report of the Railroad Commissioner*, 1890, p. xvi.
2. Jack and Jacobs, *History of Roanoke County*, p. 92–94.
3. "The Celebrated Dyspepsia Water!" *Times* (Richmond, VA), April 13, 1890.
4. See note 1 above.
5. *Report of the Postmaster-General*, 1893, p. 417.
6. "For the City of Mexico and Intermediate Points," *Sun* (Baltimore, MD), April 20, 1889.

3. Souls

1. Pattie Love Carrington to Ann Adele Hutcheson Brown, January 11, 1889, Hutcheson and Allied Families Papers, 1837–1997, MS 496, Woodson Research Center, Fondren Library, Rice University.
2. "Railway Mail Crane," Nancy A. Pope, National Postal Museum, December 2007, https://postalmuseum.si.edu/collections/object-spotlight/railway-mail-crane.
3. "Stray Thoughts," *Shenandoah Herald* (Woodstock, VA), July 19, 1889.
4. "Baum's Popular Shopping Place," *Evening Star* (Washington, DC), May 13, 1889.
5. "Two of Our Men Lost," *Richmond (VA) Dispatch*, July 6, 1889.
6. Holderness, *Reminiscences of a Pullman Conductor*, p. 16.

7. William Allen Carrington to Alice Young, January 9, 1885, Hutcheson and Allied Families Papers, 1837–1997, MS 496, Woodson Research Center, Fondren Library, Rice University.
8. "A Nine Days' Wonder!" *Knoxville (TN) Journal*, March 17, 1889.
9. Harris, *A Prince in Israel*, p. 49.
10. Ibid., p.81.
11. Ibid., p. 149.

4. Forewarning

1. *Report of the Railroad Commissioner*, 1890, p. xii.
2. Testimony of L. P. Summers, *Marshall's Administrator v. Norfolk & Western*, 1893.
3. Ibid.

5. Disaster

1. Folsom, "Railroad Washouts," p. 307.
2. Testimony of Stephen Hurt, *Marshall's Administrator v. Norfolk & Western*, 1893.
3. Ibid.
4. *Report of the Railroad Commissioner*, 1890, p. x.
5. Testimony of L. P. Summers, *Marshall's Administrator v. Norfolk & Western*, 1893.
6. Ibid.
7. Ibid.
8. *Report of the Railroad Commissioner*, 1890, p. xx.
9. Ibid.
10. Ibid.
11. Ibid.
12. Ibid., p. x.
13. Ibid.
14. "I am a Virginian," *Dallas Morning News*, July 27, 1889.
15. "Terrible Accident," *Roanoke (VA) Daily Times*, July 3, 1889.
16. Ibid.
17. *Report of the Railroad Commissioner*, 1890, p. xx.
18. Ibid.
19. "That Awful Wreck," *Daily Virginian* (Lynchburg, VA), July 6, 1889.
20. *Report of the Railroad Commissioner*, 1890, p. xxi.

6. Conflagration

1. "The Story of the Wreck," *Washington Post*, July 4, 1889.
2. *Report of the Railroad Commissioner*, 1890, p. xx.
3. "The Railroad Wreck," *Daily News* (Lynchburg, VA), July 4, 1889.
4. "The Big Wreck," *Cleveland (TN) Weekly Herald*, July 11, 1889.
5. See note 3 above.

7. Aftermath

1. Stuart, *Sermons*, p. 63–64.
2. *Report of the Railroad Commissioner*, 1890, p. xxxix.
3. Ibid., p. xxxvii.
4. Ibid., p. xxvi.
5. "The Water's Work," *Richmond (VA) Dispatch*, July 5, 1889.
6. "Scenes of Terror," *Richmond (VA) Dispatch*, July 4, 1889.
7. "An Awful Calamity," *Cleveland (TN) Weekly Herald*, July 4, 1889.
8. "The Railroad Disaster," *Daily News* (Lynchburg, VA), July 3, 1889.
9. Ibid.
10. "That Deadly Wreck," *Daily Times* (Richmond, VA), July 6, 1889.
11. "Mr. John Stevenson," *Richmond (VA) Dispatch*, July 5, 1889.
12. See note 8 above.
13. *Report of the Railroad Commissioner*, 1890, p. xliii.
14. "Two of Our Men Lost," *Richmond (VA) Dispatch*, July 6, 1889.
15. Ibid.
16. "The Thaxton Wreck," *Dallas Morning News*, July 4, 1889.
17. "Worse and Worse," *Daily Virginian* (Lynchburg, VA), July 4, 1889.
18. See note 14 above.
19. Ibid.
20. See note 7 above.
21. Stuart, *Sermons*, p. 61–62.
22. Ibid., p. 62.
23. Ibid.
24. Ibid., p. 63.
25. Ibid.
26. Ibid.
27. Ibid.
28. Harris, *A Prince in Israel*, p. 187.
29. Ibid.

8. Blame

1. "The Horror at Thaxton," *Washington Post*, July 30, 1889.
2. James C. Cassell to W. H. Ford, July 17, 1889, *Ford v. Norfolk & Western*, 1895.
3. Ibid.
4. James C. Cassell to W. H. Ford, July 29, 1889, *Ford v. Norfolk & Western*, 1895.
5. Ibid.
6. Mary M. Ford to James C. Cassell, August 1, 1889, *Ford v. Norfolk & Western*, 1895.
7. "Two of Our Men Lost," *Richmond (VA) Dispatch*, July 6, 1889.
8. "The So-Called Accident on the Norfolk and Western Railroad," *Richmond (VA) Dispatch*, July 9, 1889.
9. Ibid.
10. "A Sort of Fourth of July Pot-Pie," *Free Lance* (Fredericksburg, VA), July 9, 1889.
11. Ibid.

12. "The Norfolk and Western Railroad Holocaust," *Free Lance* (Fredericksburg, VA), July 9, 1889.
13. See note 8 above.
14. See note 12 above.
15. "Worse and Worse," *Daily Virginian* (Lynchburg, VA), July 4, 1889.
16. Testimony of L. P. Summers, *Marshall's Administrator v. Norfolk & Western*, 1893.
17. *Report of the Railroad Commissioner*, 1890, p. xlii.
18. "Thirty-Five Persons Supposed to Have Been Lost in the Wreck," *Daily Times* (Richmond, VA), July 4, 1889.
19. "A View of the Wreck," *Richmond (VA) Dispatch*, July 4, 1889.
20. "The Story of the Wreck," *Washington Post*, July 4, 1889.
21. "The Norfolk and Western Wreck," *Alexandria (VA) Gazette*, July 5, 1889.

9. Who Is John Dowell?

1. *Report of the Railroad Commissioner*, 1890, p. xxviii.
2. *Dowell's Administrator v. Norfolk & Western*, 1890.
3. "Around in Georgia," *Augusta (GA) Chronicle*, August 22, 1889.
4. Ibid.
5. "Two of Our Men Lost," *Richmond (VA) Dispatch*, July 6, 1889.
6. "Gathered at Lynchburg," *Richmond (VA) Dispatch*, July 6, 1889.
7. Ibid.
8. "Official List," *Knoxville (TN) Journal*, July 4, 1889.

10. Down the Line

1. "Held Throttle of the Flyer," *Times* (Richmond, VA), May 6, 1900.
2. "Everything But Death," *Knoxville (TN) Journal*, July 5, 1889.
3. "The Big Wreck," *Cleveland (TN) Weekly Herald*, July 11, 1889.
4. "Terrible Accident," *Roanoke (VA) Daily Times*, July 3, 1889.
5. "The Great Wreck," *Cleveland (TN) Weekly Herald*, July 25, 1889.
6. See note 1 above.
7. "Class N Locomotives" (unpublished spreadsheet provided by Jim Blackstock, Norfolk & Western Historical Society, 2012), Excel file.
8. Rebecca Jackson-Clause, "Thaxton Train Wreck Was One of the Worst," *Bedford (VA) Bulletin*, May 15, 1996.
9. Ibid.

11. Sacred Dust

1. "Memorial Services," *Cleveland (TN) Weekly Herald*, July 11, 1889.
2. "An Awful Calamity," *Cleveland (TN) Weekly Herald*, July 4, 1889.

12. Biography

1. "Burton Marye, 62, Expires at Bon Air," *Richmond (VA) Times-Dispatch*, December 6, 1924.
2. "Thirty Killed," *Daily Virginian* (Lynchburg, VA), July 3, 1889.
3. "L. P. Summers Is Held in Virginia," *Winston-Salem (NC) Journal*, December 5, 1924.
4. Randy L. Goss, e-mail message to author, September 5, 2012.
5. "That Deadly Wreck," *Daily Times* (Richmond, VA), July 6, 1889.
6. "Terrible Accident," *Roanoke (VA) Daily Times*, July 3, 1889.
7. Deposition of Dr. S. N. Jordan, October 28, 1890, *Dexter v. Norfolk & Western*, 1895.
8. Ibid.
9. Medical Notes for Robert B. Goodfellow, File 10076, Box 161, Records of St. Elizabeths Hospital, Record Group 418, National Archives Building, Washington, DC.
10. Riggs, *The Day Cleveland Cried*, p. 42.
11. Ibid., p. 43.
12. Ibid., p. 49.
13. "Tribute of Respect," *Daily News* (Lynchburg, VA), July 7, 1889.
14. *Report of the Railroad Commissioner*, 1890, p. xliii.
15. Ibid.
16. Ibid.
17. Peyton Society of Virginia, *The Peytons of Virginia*, p. 696.
18. "Virginia, Births and Christenings, 1853–1917," index, FamilySearch (https://familysearch.org/pal:/MM9.1.1/VR5X-7J1 : accessed September 8, 2012), Chas. L. Peyton in entry for Charline Peyton, 06 Dec 1888.
19. "Louis Leon Hacquard -Wheelersburg Cemetery–Wheelersburg, OH–Zinc Headstones on Waymarking.com," Waymarking.com, accessed September 9, 2012, http://www.waymarking.com/waymarks/WM1R6H_Louis_Leon_Hacquard_Wheelersburg_Cemetery_Wheelersburg_OH.
20. "A Survivor in Chattanooga," *Chattanooga (TN) Evening News*, July 4, 1889.
21. Ibid.
22. "I am a Virginian," *Dallas Morning News*, July 27, 1889.
23. "A Model Woman," *Salem Times-Register* (Salem, VA), July 5, 1889.
24. "The Big Wreck," *Cleveland (TN) Weekly Herald*, July 11, 1889.
25. Ibid.
26. "Edmund L. DuBarry," *Times Dispatch* (Richmond, VA), December 5, 1908.
27. *Report of the Railroad Commissioner*, 1890, p. xlv.
28. "Local News," *Alexandria (VA) Gazette*, October 26, 1875.
29. *Report of the Railroad Commissioner*, 1890, p. xliv.
30. Deposition of John T. Rowntree, November 20, 1890, *John T. Rowntree v. Norfolk & Western*, 1895.
31. See note 29 above.
32. See note 27 above.
33. Ibid.
34. "A Nine Days' Wonder!" *Knoxville (TN) Journal*, March 17, 1889.
35. "Death in a Wreck," *Washington Post*, July 3, 1889.
36. "The Late Disaster," *Roanoke (VA) Daily Herald*, July 4, 1889.

37. William Allen Carrington to Alice Young, January 18, 1880, Hutcheson and Allied Families Papers, 1837–1997, MS 496, Woodson Research Center, Fondren Library, Rice University.
38. Joseph Chappell Hutcheson to Elise Hutcheson Chapin, July 11, 1889, private collection of Jim Hutcheson.
39. Ibid.
40. Nina Wilson to Collins Denny, July 10, 1889, Papers of Collins Denny, Accession #2672, Special Collections Dept., University of Virginia Library, Charlottesville, VA.
41. Nina Wilson to Collins Denny, August 15, 1889, Papers of Collins Denny, Accession #2672, Special Collections Dept., University of Virginia Library, Charlottesville, VA.
42. Harris, *A Prince in Israel*, p. 209.
43. "For the City of Mexico and Intermediate Points," *Sun* (Baltimore, MD), April 20, 1889.
44. *Sketch Book of Lynchburg*, p. 128.
45. "All Had Narrow Escapes," *Sun* (Baltimore, MD), July 5, 1889.
46. Ayers, "Hon. H. S. K. Morison," p. 767.

BIBLIOGRAPHY

Ancestry.com. U.S. City Directories, 1821–1989 (Beta) [database online]. Provo, UT, USA: Ancestry.com Operations, Inc., 2011.

Ayers, H. J. "Hon. H. S. K. Morison." Edited by George Bryan. *Virginia Law Register* (J. P. Bell Company) IX (1904): 763–770.

Baldwin Locomotive Works Engine Specifications, 1869–1938, Volume 13. Southern Methodist University, Central University Libraries, DeGolyer Library.

Barger, Ralph L. *A Century of Pullman Cars.* Vol. 1, *Alphabetical List.* Sykesville, MD: Greenberg Publishing Company, 1988.

Barger, Ralph L. *A Century of Pullman Cars.* Vol. 2, *The Palace Cars.* Sykesville, MD: Greenberg Publishing Company, 1990.

Butt, Israel L. *History of African Methodism in Virginia or Four Decades in the Old Dominion.* Hampton, VA: Hampton Institute Press, 1908.

Christian, W. Asbury. *Richmond, Her Past and Present.* Richmond, VA: L. H. Jenkins, 1912.

Denny, Collins. Papers. Albert and Shirley Small Special Collections Library. University of Virginia Library.

Dexter v. Norfolk & Western Railroad Company. (Bedford County Circuit Court, 1895).

Dowell's Administrator v. Norfolk & Western Railroad Company. (Bedford County Circuit Court, 1890).

Folsom, Charles W. "Railroad Washouts." *Journal of the Association of Engineering Societies* (The Board of Managers of the Association of Engineering Societies) 5, no. 8 (June 1886): 307.

Ford v. Norfolk & Western Railroad Company. (Bedford County Circuit Court, 1895).

Fourteenth Annual Report of the Railroad Commissioner of the State of Virginia. Richmond, VA: J. H. O'Bannon Superintendent of Public Printing, 1890.

Harris, Carlton Danner. *Alpheus W. Wilson, a Prince in Israel.* Louisville, KY: Board of Church Extension of the Methodist Episcopal Church, South, 1917.

Holderness, Herbert O. *The Reminiscences of a Pullman Conductor or Character Sketches of Life in a Pullman Car.* Chicago: s.n., 1901.

Huntington, William S., and Charles Latimer. *The Road-Master's Assistant and Section-Master's Guide.* Sixth. New York: The Railroad Gazette, 1881.

Hutcheson and Allied Families. Papers. Woodson Research Center, Fondren Library, Rice University.

Jack, George S., and Edward Boyle Jacobs. *History of Roanoke County.* Roanoke, VA: Stone, 1912.

Jefferson, Thomas. *Notes on the State of Virginia.* London, 1787.

John T. Rowntree v. Norfolk & Western Railroad Company. (Bedford County Circuit Court, 1895).

Kappa Sigma. *Caduceus of Kappa Sigma.* Vol. XX. Nashville, TN, 1905.

Marshall's Administrator v. Norfolk & Western Railroad Company. (Bedford County Circuit Court, June 2, 1893).

McGroarty, John Steven. *Los Angeles from the Mountains to the Sea.* Vol. II. Chicago and New York: The American Historical Society, 1921.

Murray, Major J. Ogden. *The Immortal Six Hundred: A Story of Cruelty to Confederate Prisoners of War.* Winchester, VA: The Eddy Press Corporation, 1905.

"Norfolk & W. R. Co. v. Marshall's Adm'r." *The Southeastern Reporter* (West Publishing Co.) 20 (1895): 823–824.

Norfolk & Western R. R. Co. *Virginia Summer Resorts.* Buffalo, NY: Matthews, Northrup & Co., 1889.

Peaks of Otter Chapter, Daughters of the American Revolution. *Bedford Villages—Lost and Found.* Vol. 3. Bedford, VA: Peaks of Otter Chapter, Daughters of the American Revolution, 2000.

Peyton Society of Virginia. *The Peytons of Virginia II.* Vol. Two. Gateway Press, 2004.

Pullman Company Employee Records. The Newberry Library, Chicago.

Records of St. Elizabeths Hospital. Record Group 418. National Archives Building, Washington, DC.

Report of the Postmaster-General of the United States. Washington, DC: Government Printing Office, 1893.

Report of the Postmaster-General of the United States. Washington, DC: Government Printing Office, 1890.

Reutter, Mark, ed. "Shaw's All-Time List of Notable Railroad Accidents, 1831–2000." *Railroad History* (The Railway & Locomotive Historical Society), no. 184 (Spring 2001): 37–45.

Riggs, Debbie. *The Day Cleveland Cried.* Cleveland, TN: Debbie Riggs, 2011.

Sketch Book of Lynchburg, VA. Its People and its Trade. Lynchburg, VA: Edward Pollock and S. C. Judson, 1887.

Stuart, George R. *Sermons.* Philadelphia: Pepper Publishing Company, 1904.

"The Railroad Roll." *Travelers Record* (Travelers Insurance Co.) XXV, no. 5 (August 1889): 6.

Thirty-Second Annual Reunion of the Association of Graduates of the United States Military Academy at West Point. Saginaw, MI: Seemann & Peters, 1901.

Tyler, Lyon G., ed. *Men of Mark in Virginia: Ideals of American Life.* Vol. III. Washington, DC: Men of Mark Publishing Company, 1907.

Tyler, Lyon G., ed. *Men of Mark in Virginia: Ideals of American Life.* Vol. IV. Washington, DC: Men of Mark Publishing Company, 1908.

ABOUT THE AUTHOR

Michael Jones was planted in West Virginia and cultivated in South Carolina, with roots extended deep into Virginia soil. He is a graduate of Clemson University, author, entrepreneur, master naturalist, and history lover. You can find Michael online at mikejoneswrites.com.

facebook.com/mikejoneswrites
twitter.com/mikejoneswrites
instagram.com/mikejoneswrites

Made in the USA
Columbia, SC
07 July 2025